HUMAN GENETICS

HUMAN GENETICS

SIXTH EDITION

SD Gangane, MBBS, MS (ANATOMY), FAIMS
Professor
Department of Anatomy
Terna Medical College, Nerul
Navi Mumbai, Maharashtra, India

Ex-Professor and Head
Genetic Division and
Department of Anatomy
Grant Medical College
Mumbai, Maharashtra

Ex-Professor and Head
Department of Anatomy
RCSM Government Medical College
Kolhapur, Maharashtra

ELSEVIER

ELSEVIER

RELX India Pvt. Ltd.
Registered Ofice: 818, 8th Floor, Indraprakash Building, 21, Barakhamba Road, New Delhi 110001
Corporate Ofice: 14th Floor, Building No. 10B, DLF Cyber City, Phase II, Gurgaon-122002, Haryana, India

Human Genetics, 6e, SD Gangane

ISBN: 978-81-312-6209-2
e-book ISBN: 978-81-312-6210-8

Content Strategist - Education Solutions: Arvind Koul
Content Project Manager: Goldy Bhatnagar
Sr Production Executive: Dhan Singh
Sr Graphic Designer: Gopalakrishnan Venkatraman

Typeset by GW India

Printed and bound at Rajkamal Electric Press, Kundli, Haryanan.

Dedicated to

Satguru Sant Shri Gajanan Maharaj

Preface to the Sixth Edition

Genetics makes the most dynamic sector in Medical World. Everyday dawns with some quantum leap in existing knowledge of various diseases – genetic as well as non-genetic, as they were thought to be once upon a time. This refers to common diseases like diabetes, hypertension, obesity, coronary artery disease, Alzheimer disease, etc. However, later it was discovered that all these common diseases do have genetic component. In short, practically every disease has genetic component playing a significant role. All these facts make it obligatory to put all the new developments in this sixth edition.

Among recent advances the CRISPR-Cas System, as a gene editing tool has been made a significant impact in the field of gene therapy. It also seems to be cost effective as compared to earlier known methods of gene therapy. Epigenetics also has gathered momentum in recent times. Many researchers are working in this field to find ways and means by which better healthcare can be provided. It will also help in better understanding of pathophysiology of the disease process and help in prevention of many diseases. Keeping this in mind a new chapter on Epigenetics has been included.

The competency-based medical education (CBME) has been advocated by National Medical Commission and also stressed the importance of genetics. The aim of this book is to make some basic understanding of genetics essential for Indian Medical Graduates. In this context the competency codes integrated within the text as per new competency-based undergraduate curriculum. The competency map also has been given in the beginning of the book to facilitate localization of various competencies in genetics.

Suggestions from both, the faculty and students are solicited to enable me to improvise this title in subsequent editions. With this I humbly submit this edition to the readers.

SD Gangane

Preface to the First Edition

Human genetics claims a frontline position under the faculty of medicine. It is duly emphasised if we realise the wide spectrum of patients referred to this speciality. The medical fraternity is becoming aware of present day techniques in this rapidly expanding branch of medicine. The vigilant eye of a family physician can promptly put, the family at risk, to the care of the geneticist. In fact he serves as a liaison between the patient, his relatives and the geneticist. It is from this point of view that "Genetics" forms a part of undergraduate medical curriculum in most of the universities.

Genetics makes almost a boundless field and every branch of genetics is vast enough to perplex a new entrant to this field. A simple and concise account elucidating human genetics is therefore highly desired and always welcome. This book forms a concise version chiefly designed to cater to the needs of an undergraduate student in a medical school/college. Schematic representations along with clinical photographs of patients have been incorporated to simplify and supplement the text. The subject has been dealt with in 12 chapters. The first chapter presents historical gleanings followed by other chapters containing cytogenetics, molecular genetics, biochemical genetics, immunogenetics and so on. The aim has been to offer basic principles without superfluous details.

This book has taken its present form with direct or indirect help from many people. I sincerely thank them. My special thanks to Mrs. N.N. Bhagat who had been of great help in typing the manuscript, Dr. A.L. Kulkarni, Associate Professor of Anatomy for his assistance in preparing the manuscript. I am grateful to Miss Vidya Dicholkar and Mr. More from Genetic Division, Grant Medical College, Mumbai for their help in photomicrography and clinical photography.

I am also grateful to Dr. (Mrs.) A.M. Lete, Professor of Anatomy, Grant Medical College, Mumbai for the constant encouragement and valued suggestions that I got during the writing of this book. With this, I humbly submit this book to the readers.

SD Gangane

Acknowledgements

With deep sense of gratitude, I wish to acknowledge the help rendered by members of the Genetic Division, Department of Anatomy, Grant Government Medical College and Sir JJ Group of Hospitals, Mumbai. I am also grateful to Dr DS Joshi, Professor and Head, Department of Anatomy, GMC, Mumbai, for his valuable suggestions and constant support.

A legendary figure in the field of Anatomy, Professor Vishram Singh has been an inspiration for me and many people like me. He has been a guiding force for budding anatomists. Sir, I am grateful to you for encouragement and guidance.

During COVID-19 lockdown period, the sixth edition has surfaced through the cooperation and coordination of two persons to whom I am grateful – Mr Shantanu Gangane (my son) and Dr Shabana Borate my ex-colleague and Associate Professor at Grant Government Medical College, Mumbai. Special thanks to Dr Shabana for her untiring and immense technical support in making of this manuscript of sixth edition.

The book in its present form has seen the light of the day because of persistent efforts of the entire team of RELX India Pvt. Ltd. especially Arvind Koul (Content Strategist), Shabina Nasim (Head of Content Project Management) and Goldy Bhatnagar (Content Project Manager).

Contents

Competency Map

Competency Code	Competency	Core (Y/N)	Chapter Nos.	Page No.
Topic: Chromosomes				
AN 73.1	Describe the structure of chromosomes with classification	Y	2	16–18
AN 73.2	Describe the technique of karyotyping with its applications	Y	2	17–21, 23
AN 73.3	Describe the Lyon's hypothesis	Y	2	24–26
Topic: Patterns of Inheritance				
AN 74.1	Describe the various modes of inheritance with examples	Y	6	96–105, 106–108, 109, 110
AN 74.2	Draw pedigree charts for the various types of inheritance and give examples of diseases of each mode of inheritance	Y	6	97, 98–99, 103–105, 106–108, 109
AN 74.3	Describe the multifactorial inheritance with examples	Y	6	112–113
AN 74.4	Describe the genetic basis and clinical features of achondroplasia, cystic fibrosis, vitamin D resistant rickets, haemophilia, Duchene's muscular dystrophy and sickle cell anaemia	N	6 7	100, 105, 108, 109, 124–125

Continued

Competency Code	Competency	Core (Y/N)	Chapter Nos.	Page No.
Topic: Principle of Genetics, Chromosomal Aberrations and Clinical Genetics				
AN 75.1	Describe the structural and numerical chromosomal aberrations	Y	4	65–66, 70, 74–75
AN 75.2	Explain the terms mosaics and chimeras with example	N	2 4	26 82, 83
AN 75.3	Describe the genetic basis and clinical features of Prader–Willi syndrome, Edward syndrome and Patau syndrome	N	4	66–68, 73, 74
AN 75.4	Describe the genetic basis of variation: polymorphism and mutation	Y	8 3	138 42
AN 75.5	Describe the principles of genetic counselling	Y	14	205, 206, 210–214
Topic: Fetal Membranes				
AN 80.4	Describe the embryological basis of twinning in monozygotic and dizygotic twins	Y	12	189–191
Topic: Prenatal Diagnosis				
AN 81.1	Describe the various methods of prenatal diagnosis	Y	13	198, 199, 201–203
AN 81.2	Describe the indications, process and disadvantages of amniocentesis	Y	13	198–199
AN 81.3	Describe the indications, process and disadvantages of chorion villus biopsy	Y	13	199

Historical Gleanings

1

LEARNING OBJECTIVES

At the end of this chapter the students should be able to understand:

- How the science of Genetics evolved?
- Salient events in its evolution chronologically
- Principles of Mendelism
- Recent developments in the field of Genetics

KEY WORDS

Mendel, Mendelism

History is the documentation of facts or events in the past. It facilitates us to learn about present and plan for the future. As far as Genetics is concerned, the concept of heredity dates back around 6000 years, as evidenced in the stone engravings from Chaldea in Babylonia in current Iraq. These engravings depict pedigrees relating to some characteristics in horses. As regards human heredity, haemophilia (a bleeding disorder) was the first genetic disorder known about 1500 years ago. Until recent times, however, a proper record of the historical events, the nature of the hereditary disorders and explanations regarding their etiology remained largely speculative.

Mendel

(*Source:* Emery's Elements of Medical Genetics, Fig. 1.1, 2012, Churchill Livingstone.)

In the third century BC, Aristotle suggested that semen originates from blood and has an ability to give life to embryo; embryo is formed in uterus by coagulation of menstrual blood. This concept ruled for over 2000 years. In the 17th century, Dutch scientist Regnier de Graaf demonstrated that union of egg (female germ cell) and sperm (male germ cell) is essential for conception. He described small protuberances on mammalian ovaries, now called Graafian follicles, containing unfertilised egg. This led to a thought that the sperm alone is not responsible as a sole hereditary agent.

Pierre-Louis Moreau de Maupertuis, a French naturalist born in 1698 in France, studied polydactyly (extra digit) and albinism (lack of pigment). He showed that these two traits were inherited in different ways. He proposed that there were hereditary particles responsible for the formation of a particular body part. Each body part was formed by two such particles, one from each parent. One particle may dominate over the other (recessive). This was much on the lines of what Mendel propounded a century later. He believed that both parents contribute equally to their offsprings. He substantiated his claim by animal experiments.

A scientific approach towards genetics came in the later half of 19th century when Gregor Mendel gave principles of heredity. Around the same period, in 1859, Charles Darwin postulated in his "Theory of Evolution" that multiplication of individuals of a given species is associated with the origin of variations resulting from recombination and mutation.

MENDEL AND MENDELISM

Our present knowledge of genetics has its roots in Mendelian principles—that is where we start with. Johann Mendel was born on 22 July 1822 in Heinzendorf in Moravia situated in old time Austria, now recognised as Czechoslovakia. He adopted the name "Gregor" in 1843 and subsequently became a priest. In 1851, he joined the University of Vienna. Here he was greatly influenced by two scientists—Franz Unger, a plant physiologist, and Christian Doppler, after whom the Doppler effect in physics is named. A part of Unger's teaching course incorporated plant hybridisation experiments by two German botanists, Kölreuter and Gaertner. In fact Gaertner worked on peas, the same material that Mendel used about a decade later. Gaertner, however, could not interpret the results successfully.

In 1853, Mendel went to Brunn where he conducted his experiments on garden peas (*Pisum sativum*). He selected seven pairs of contrasting characters in the garden pea, such as height of plant, shape of pod, texture of seed, flower position and colour, etc. (Fig. 1.1).

He crossed these varieties of plants considering one pair of contrasting character. Hybrids thus obtained formed F_1 generation. Plants in this (F_1) generation were allowed to undergo self-pollination. This led to F_2 generation. The plants in F_1 generation resembled one of the parents; for example, cross between tall and dwarf resulted in all tall plants. The characteristics expressed in hybrids were called dominant, e.g. tallness. Among the two characters—tall and dwarf—tallness expressed itself and so was called a dominant character. The characters that were not manifested in hybrids were referred to as recessive ones. Analysis of F_2 generation revealed both types of plants, one expressing dominant character (tall) and the other expressing recessive character (dwarf). They occurred in ratio 3 tall:1 dwarf. F_2 generation plants exhibiting recessive character were self-pollinated. This resulted in F_3 generation with all the plants expressing recessive character, i.e. dwarf. Plants expressing dominant character (tall) in F_2 generation on self-pollination yielded

Figure 1.1 Seven contrasting characters in garden peas.

plants of two types. Two-thirds of them on self-pollination resulted in plants expressing both dominant and recessive characters, i.e. tall and dwarf. Rest one-third on self-pollination displayed only dominant character. This led to the conclusion that F_2 generation is constituted by 1:2:1 ratio of plants, the analysis being based on the type of offsprings they produce.

Let us assume two contrasting characters, tall represented by letter "T" and dwarf by letter "t". The former represents a dominant character, while the latter is recessive. The result of their progeny in successive generations is shown in Fig. 1.2.

Mendel's Laws

On the analysis of progeny of generations in garden pea, Mendel propounded his concepts that came to be recognised as Mendel's Laws:

1. **Law of Unit Inheritance**
   Under modern teaching in genetics, this concept is hardly stressed. In pre-Mendelian era, concept about inheritance was that the characteristics of the parents blend in the offsprings. Mendel, for the first time, offered a

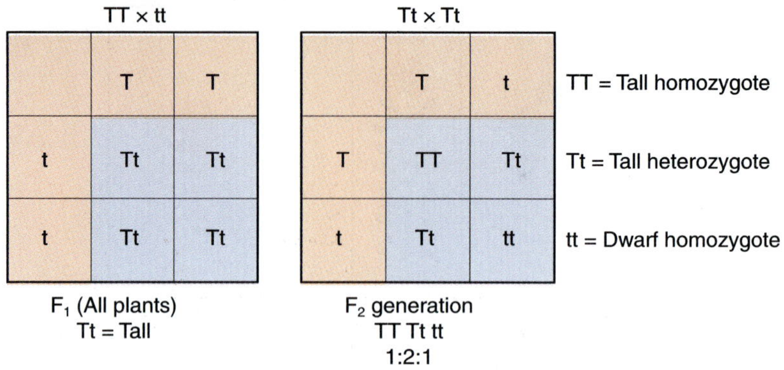

Figure 1.2 Progeny of tall and dwarf plant (homozygous) in F_1 followed by self-pollination of F_1 giving rise to F_2 generation.

new concept that characters do not blend; if they do not express in the first generation, they can reappear without change in the subsequent generation. For example, we have seen that the cross between tall and dwarf plants led to F_1 generation having all tall plants. The dwarfness reappeared in F_2 generation. There was no blending of characters like tall + dwarf = medium-sized plant.

2. **Law of Segregation**
   This law states that the members of a gene pair segregate and pass to different gametes. They are never found in the same gamete, except in the event of non-disjunction, i.e. when members of chromosome pair fail to separate (Fig. 1.3). This law applies to the genes on homologous chromosomes, and disjunction of paired homologous chromosomes forms the basis of the law of segregation.
3. **Law of Independent Assortment**
   It states that the members of different gene pairs assort independently of one another during gametogenesis. The law of independent assortment applies to the behaviour of non-homologous chromosomes in the meiosis. The genes located on these non-homologous chromosomes

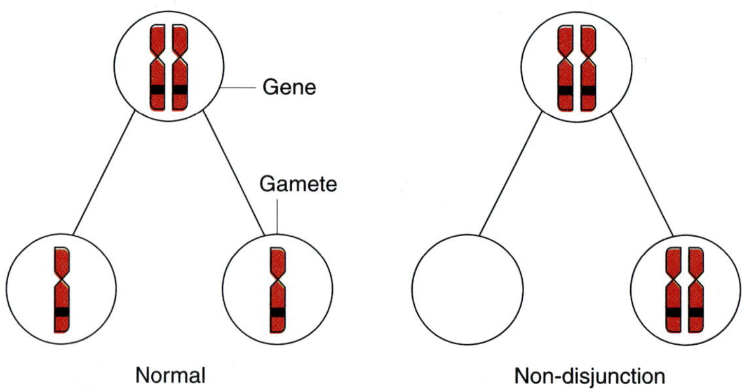

Figure 1.3 Law of segregation.

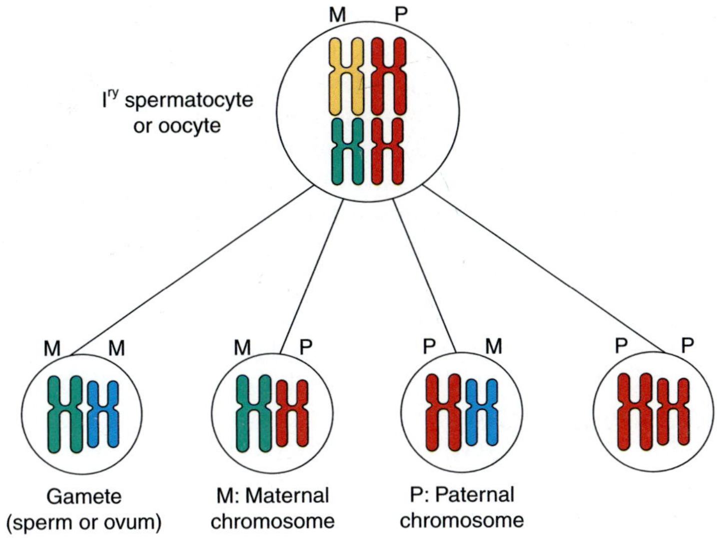

Figure 1.4 Law of independent assortment.

(i.e. genes that are not linked) undergo independent assortment. Random assortment of maternal and paternal chromosomes forms a physical basis of the law of independent assortment (Fig. 1.4).

Mendel presented the results of his work in 1865 before the Natural History Society of Brunn. In the subsequent year, it was published in the *Transactions of the Society*, not much widely read. In fact, Mendel's work remained buried in the pages of history till the turn of the century, for almost 35 years. In 1900, Mendelism was rediscovered by three botanists independently, namely Professor Hugo de Vries from Amsterdam, C Correns from Tübingen, and Erich von Tschermak, an agricultural assistant from Esslingen near Vienna. It is unfortunate that Mendel's work saw the light of the day 16 years after his death. Mendel died of Bright disease in 1884.

MILESTONES IN THE DEVELOPMENT OF GENETICS

Karl Landsteiner discovered ABO blood groups in the year 1900. Walter S. Sutton and Theodor Boveri independently proposed chromosomal theory of heredity in 1903. In 1902 William Bateson, a strong proponent of Mendelism, coined the term "Genetics". A Danish botanist, Johannsen, introduced the term "Gene" in 1909.

The first example of Mendelian inheritance in man was reported in 1902 by Garrod. It was a case of alkaptonuria. A few years later in 1908, Garrod coined up his concept of "Inborn errors of metabolism". Bridges (1916) demonstrated that genes are sequences of nucleotides and they are oriented in a linear fashion on chromosomes. This was the beginning of cytogenetics.

In 1927, Muller showed that X-ray exposure increases the mutation rate. The effect of certain other extraneous factors is also similar. They include ultraviolet (UV) rays, cosmic rays, gamma rays, certain drugs and dyes.

The concept of one gene–one enzyme was proposed by Beadle and Tatum in 1941. Barr and Bertram (1949) demonstrated "Barr body" in female cat neurons. In 1952, Gerty Cori and Carl Cori demonstrated an enzyme defect in glucose-6-phosphate-dehydrogenase (G6PD) deficiency, a type of inborn error of metabolism.

The double helical model of DNA molecule was forwarded by Watson, Crick and Wilkins in 1953, for which they were awarded the Nobel Prize. A major breakthrough came in 1956 when JH Tjio and A Levan, and independently Ford and Hamerton demonstrated the actual number of human chromosome complement as "46"; earlier it was thought to be 48. This was made possible by improved cytogenetic techniques evolved by these scientists.

It was in 1959 that Lejeune and his associates noted for the first time a chromosomal defect as the reason of Down syndrome. During the same period, a specific chromosomal aberration associated with cancer was detected by Nowell and Hungerford. This was labelled as Philadelphia chromosome. It is found in patients of chronic myeloid leukaemia. In 1961, Mary F. Lyon proposed the hypothesis of X chromosome inactivation in female cells. In 1970, Har Gobind Khorana and his associates succeeded in synthesising a gene de novo. It was non-functional, though structurally correct. But by 1976, Khorana and his colleagues were able to synthesise a functional artificial gene.

A review of the last three–four decades duly stresses the importance of genetics, if we consider the names of Nobel Laureates in physiology and medicine. Here is the list of these legends. In 1962, Francis Crick, James D. Watson and Maurice Wilkins got the Nobel Prize for discovering the molecular structure of DNA. Francis Jacob, J Monod and A Lwoff won the Nobel Prize for their work on regulation of gene in 1965. Deciphering of the genetic code won the Nobel Prize for Robert Holley, Har Gobind Khorana and Marshall Nirenberg in 1968. In 1975, R Dulbecco, H Temin and D Baltimore worked out an interaction between tumour viruses and nuclear DNA. The discovery of restriction endonucleases won the Nobel Prize to W Arber, D Nathans and H Smith in 1978. In 1980, B Benacerraf, G Snell and J Dausset explained how genes control an immunological response, and they were awarded Nobel Prize for this. In 1983, Barbara McClintock received Nobel Prize for the discovery of "Jumping genes" or transposons (mobile genes). The following discoveries of cell receptors in familial hypercholesterolaemia by M Brown and J Goldstein (1985), genetic aspects of antibodies by Tonegawa Susumu (1987), the study of oncogenes by M Bishop and H Varmus (1987) won them the Nobel. In 1993, Richard Roberts and Philip Sharp offered the concept of "split genes" for which they were awarded the Nobel Prize. In 1995 Nobel Prize was awarded jointly to Edward B. Lewis, Christiane Nüsslein-Volhard and Eric F. Wieschaus for their discoveries concerning the genetic control of early embryonic development. In 2002, Sydney Brenner,

H. Robert Horvitz and John E. Sulston discovered genetic regulation of organ development and programmed cell death. Andrew Z. Fire and Craig C. Mello (in 2006) were awarded Nobel Prize jointly for their discovery of RNA interference—gene silencing by double-stranded RNA. In 2007, Nobel Prize was awarded jointly to Mario R. Capecchi, Sir Martin J. Evans and Oliver Smithies for principles for introducing specific gene modifications in mice by the use of embryonic stem cells. The Nobel Prize in 2009 was awarded jointly to Elizabeth H. Blackburn, Carol W. Greider and Jack W. Szostak for the discovery of how chromosomes are protected by telomeres and the enzyme telomerase. In 2010, Robert G. Edwards was conferred Noble Prize for the development of in vitro fertilisation.

Recent advances in the field of genetics chiefly aim at developing more accurate techniques for early and precise diagnosis of genetic disorders. Rapidly expanding areas include recombinant DNA technology, restriction fragment length polymorphisms (RFLPs), DNA fingerprinting, human genome, whole chromosome paints (WCPs), fluorescent in situ hybridisation (FISH), gene therapy and stem cell therapy.

The most striking development has been in understanding the application of CRISPER-Cas9 system, a "Gene editing tool". Concomitantly "Epigenetics" has gathered a momentum. It has revealed many facts which were unknown in the past. It has also helped us to understand that not only our DNA is to be blamed but also environmental factors have significant role in phenotype and the disease. Both CRISPER gene editing system and the understanding about epigenetics have expanded boundless opportunities for the research.

Summary

- Stone engravings from Chaldea in Babylonia (Iraq) depict pedigree of horse 6000 years ago.
- Haemophilia (bleeding disorder) is known for 1500 years ago.
- 300 BC, Aristotle suggested that semen originates from blood and has ability to give life to embryo.
- In 17th century, Regnier de Graaf demonstrated that union of egg and sperm is essential for conception. Mature ovarian follicle is named as Graafian follicle.
- In 1698, Moreu-de-Maupertuis from France studied polydactyly and albinism. He suggested that traits were inherited through hereditary particles that are received from the parents.
- Johann Mendel (1822–1884), born on 22nd July 1822, adopted name "Gregor" in 1843. He joined University of Vienna in 1851. He went to Brunn in 1853, where he conducted experiments on garden peas (*Pisum sativum*). He presented his work in 1865 before the Natural History Society of Brunn. He propounded what are recognised as Mendel's laws of inheritance—1. Laws of unit inheritance, 2. Law of segregation, 3. Law of independent assortment.
- In 1900, Karl Landsteiner discovered ABO blood group.
- In 1902, term Genetics was coined by William Bateson.

Continued

Summary—cont'd

- In 1902, the first example of Mendelian inheritance was reported by Garrod. It was a case of Alkaptonuria.
- In 1916, Bridges demonstrated that genes are sequences of nucleotides.
- In 1927, Muller demonstrated mutational effect of X-rays.
- In 1941, Beadle and Tatum gave a concept of one gene–one enzyme.
- In 1949, Barr and Bertram demonstrated "Barr body" in female cat neurons.
- In 1952, G6PD deficiency was detected by Gerty and Carl Cori.
- In 1953, double helical model of DNA molecule was given by Watson, Crick and Wilkins.
- In 1956, Lejeune noted 21 trisomy as the chromosomal error in Down syndrome.
- In 1961, Mary F. Lyon proposed the hypothesis of X inactivation.
- In 1976, Khorana and his associates synthesised functional artificial gene.
- In 1983, Barbara McClintock discovered jumping genes.
- In 1987, Varmus and Bishop studied oncogenes.
- In 1993, concept of split genes was offered by Roberts and Sharp.
- In 1995, Edward B et al. offered concept concerning the genetic control of early embryonic development.
- In 2002, Sydney Brenner et al. coined the concept of genetic regulation of organ development and programmed cell death.
- In 2006, Andrew Z. Fire and Craig C. Mello were awarded Nobel Prize for their discovery of RNA interference—gene silencing by double-stranded RNA.
- In 2007, the principles of specific gene modifications by the use of embryonic stem cells in mice were given by Mario R. Capecchi et al.
- The Nobel Prize in 2010 was awarded to Robert G. Edwards for the development of in vitro fertilisation.

QUESTION YOURSELF*

1. Mendel selected following contrasting characters in garden peas except:
   a. Height of plant
   b. Shape of pod
   c. Size of seed
   d. Texture of seed
2. In Mendelian experiment, F_2 generation had following phenotypic ratio of tall and dwarf plants:
   a. 2:1:1
   b. 1:2:1
   c. 1:1:2
   d. None of the above
3. Self-pollination of tall heterozygote plants results in progeny exhibiting:
   a. All tall homozygotes
   b. All tall heterozygotes
   c. All dwarfs homozygotes
   d. None of the above
4. What is true about law of segregation?
   a. Members of the gene pair segregate and pass to different gametes
   b. This law applies to genes on homologous chromosomes
   c. The exception to the law is an event of "non-disjunction"
   d. All of the above

*See page 257 for Answers.

Cytogenetics 2

COMPETENCY AN 73.1 Describe the structure of chromosome with classification.
COMPETENCY AN 73.2 Describe the technique of karyotyping with its applications.
COMPETENCY AN 73.3 Describe Lyon's hypothesis.
COMPETENCY AN 75.2 Explain the term mosaics with examples.

LEARNING OBJECTIVES

At the end of this chapter the students should be able to understand

- The process of mitosis and meiosis
- Morphology and classification of chromosomes
- Chromosome preparation (Karyotyping)
- Lyon's hypothesis
- Various ways to study chromosomes

KEY WORDS

Cell cycle, Mitosis, Meiosis, Chromosomes, Sex chromatin, Barr body

Cytogenetics as a diagnostic tool has taken a back seat with the advances in molecular genetics. However, it still holds a promising place in the diagnosis of genetic disorders.

Cytogenetics deals mainly with the study of chromosomes and sex chromatin. Advances in cytogenetics have made it possible to pinpoint the errors in human chromosomes. In this chapter, we shall deal with the basic facts about chromosomes and the recent techniques for studying this subject.

In eukaryotes, chromosomes reside in the nucleus. An interphase nucleus has chromosomes in coiled and extended portions. Coiled portion stains darker and appears in clumps, called *heterochromatin* (Fig. 2.1), while the extended portion is pale staining and forms *euchromatin*. Chromosome number in a particular species is fixed. The human chromosome complement consists of 46 chromosomes (23 pairs). This number 46 is called *diploid,* often designated as 2n. A *haploid* number "*n*" is 23, and it is encountered only in gametes. Of the total 46, 44 are autosomes (in 22 pairs) and 2 are

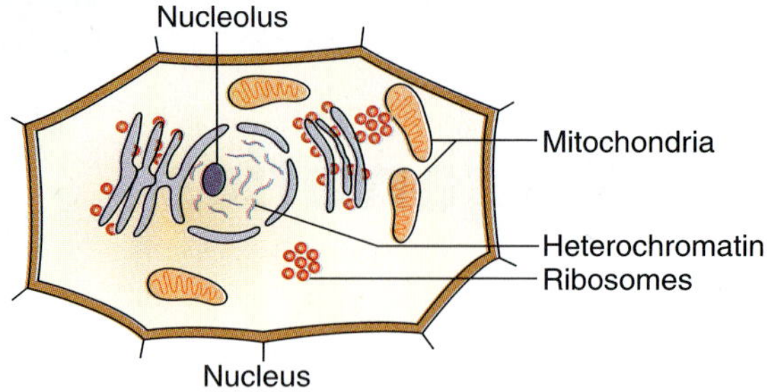

Figure 2.1 An interphase cell showing heterochromatin.

sex chromosomes. The latter in females are XX and in males are X and Y. In females, of the two X chromosomes, one is rendered inactive to form Barr body. Its significance is discussed later in this chapter.

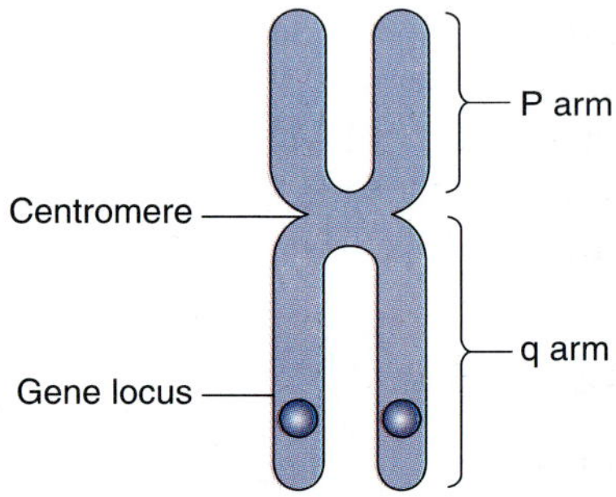

Figure 2.2 A chromosome.

The word chromosome is derived as follows: *Chroma* means colour and *soma* means body. As they appear like coloured (stained) rod-shaped structures, they are called chromosomes. Each chromosome (in metaphase) consists of two *chromatids* joined together at *centromere* (Fig. 2.2). Depending upon the placement of centromere, chromosomes are classified into various types. Chromosomes have small units of heredity called *genes,* which have specific positions on the chromosome. This is called gene *locus.* Chromosomes are usually studied in *metaphase* of mitosis. It will be interesting as well as essential to note the details of cell division so as to understand the role of chromosomes during cell division.

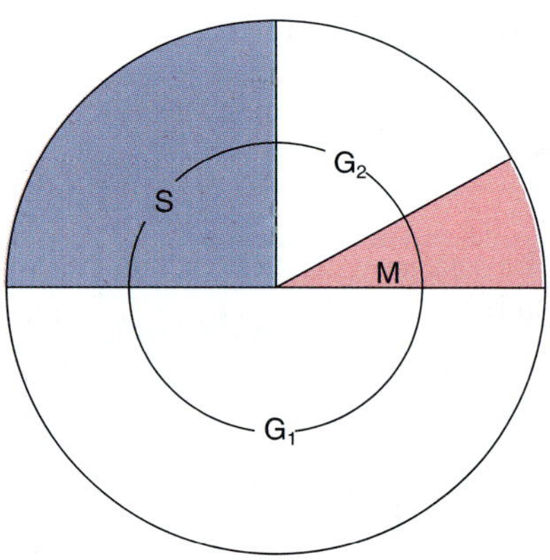

Figure 2.3 Diagram showing stages of the "cell cycle". They are: G1, resting stage of interphase; S, DNA replication; G2, second resting stage; and M, mitosis.

A cell that is capable of division undergoes a cyclical change throughout its life; this is called *cell cycle.* It consists of an interphase–mitosis–interphase cycle. Most of the body cells exist and function in an interphase. An interphase comprises the following phases: G1 (Gap 1), S (Synthesis) and G2 (Gap 2). Gap 1 phase is followed by S phase. During S phase, DNA is synthesised. It takes about 7 hours time. After this, the cell enters a brief G2 phase, which takes about half an hour. This is followed by mitosis (Fig. 2.3).

MITOSIS

It involves somatic cells. The result of mitosis is two daughter cells each with a copy of parental genome. During mitosis, both cytoplasm and nucleus divide. Cytoplasmic division is relatively a simple phenomenon, while nuclear division presents a sequence of complicated activities. For the sake of description, mitosis can be divided into four stages, namely prophase, metaphase, anaphase and telophase (Fig. 2.4).

Prophase

In this phase, the nuclear chromatin organises to form rod-like bodies called chromosomes. Each chromosome seems to be made up of two thin strands called chromatids. They are joined at the spot called centromere. Nuclear membrane disappears. Centriole duplicates itself and the two daughter centrioles move towards opposite poles.

Metaphase

Chromosomes condense further and move towards equatorial plane of the cell. They form a metaphase plate. Meanwhile microtubules radiate from centrioles to the equatorial plane. They attach to centromeres of the chromosomes. These microtubules constitute *spindle*.

Anaphase

In this phase, centromeres divide vertically and the paired chromatids disjoin. They form new daughter chromosomes. The new chromosomes move, one to each pole. The movement of chromosomes towards a pole is supposed to be

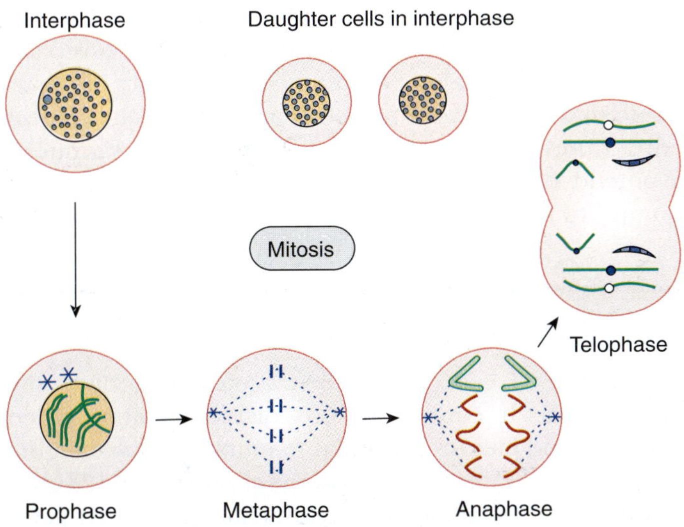

Figure 2.4 Stages of mitosis.

by contraction of spindle fibres. At this stage, an indication of cytoplasmic division appears in the form of a furrow along the equatorial plane.

Telophase

In this stage, daughter chromosomes have arrived at the poles. This is followed by *cytokinesis,* i.e. division of cytoplasm. It is accomplished by further deepening of the furrow at the equatorial plane of the cell separating two daughter cells. Each daughter cell bears identical chromosome complement. Subsequently, chromosomes start unwinding and show poor staining. Finally, they are no longer visible as separate entities, but form chromatin network. Concomitantly, there is reconstitution of nuclear membrane. Thus two daughter cells, each appearing in interphase, are obtained.

Comments

Cell division can be arrested at metaphase by substances like colchicine or its derivatives. Colchicine inhibits spindle microtubule formation. This permits us to study metaphase chromosomes.

In mitosis, two points need elaboration; these are somatic recombination and sister chromatid exchange. Somatic recombination is crossing over between homologous chromosomes in mitosis. It is less frequent than recombination in meiosis. This is because, in meiosis, homologous chromosomes are more closely associated than in mitosis, thus offering more chances of meiotic recombination.

Sister chromatid exchange

It involves crossing over between the sister chromatids of a single chromosome in mitosis. It was first demonstrated in 1957 by Taylor. Later on in 1973, a special technique was developed by Latt to demonstrate DNA replication in human metaphase chromosome. In this technique, the cultured cells are allowed to replicate twice in the presence of bromodeoxyuridine (BUdR). This allows incorporation of BUdR in newly synthesised DNA. It replaces thymine. The incorporation of BUdR alters staining characteristics of chromatids. The chromatid containing BUdR stains with fluorescent stain (Hoechst 33258). Bright and dim fluorescence pattern along chromatids signifies occurrence of sister chromatid exchange. In some genetic disorders like Bloom syndrome, frequency of sister chromatid exchanges is greatly increased.

MEIOSIS

This is a special type of cell division that occurs in gonads and results in formation of gametes. It consists of two divisions, often called meiosis I (reduction division) and meiosis II (similar to mitosis). Each daughter cell at the end of meiosis I contains haploid chromosome complement (23 chromosomes). This haploid number is maintained thereafter in (meiosis II) gametes. It is in contrast to mitosis in which diploid number is maintained in daughter cells.

Meiosis I (Reduction Division) (Fig. 2.5)

Prophase I

It is much prolonged in contrast with mitotic prophase. It consists of the following stages:

Leptotene

In this, chromosomes appear as thin thread-like structures. The chromosome threads start condensing, as a result they show alternate thick and thin

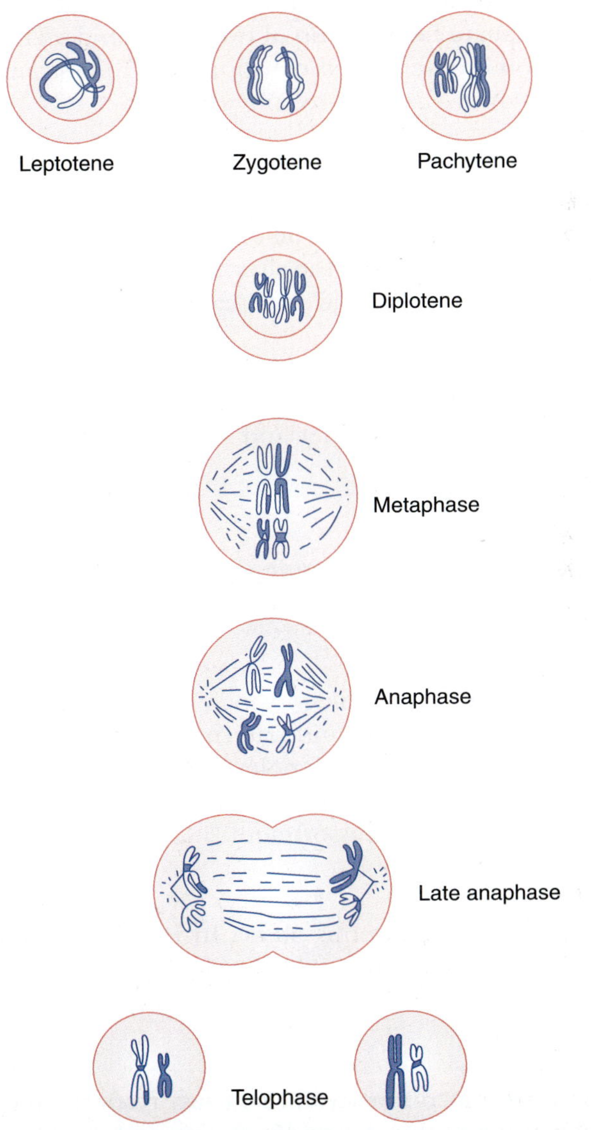

Figure 2.5 Chromosomal behaviour in meiosis. Only two pairs of chromosomes are shown. Chromosomes of one parent are shown in purple and those of other in outline. Note crossing over and exchange of material in diplotene and subsequent stages.

portions. This gives them a beaded appearance. These beads are called *chromomeres.*

Zygotene

The homologous chromosomes pair during this stage. Two members of the homologous pair lie parallel and show point-to-point pairing. Together these chromosomes are called bivalents. The pairing of homologous chromosomes occurs *only* in meiosis and not in mitosis. In sex chromosomes, pairing involves only small segments, the tips of their short arms.

Pachytene

The chromosomes become more tightly coiled and stain deeply. Each chromosome now appears to be made up of two chromatids. Thus, each bivalent (homologous chromosome pair) is constituted of four strands, hence called *tetrad.* The strands of *tetrad* show crossing at places.

Diplotene

This stage is characterised by longitudinal separation of the members of bivalent, *without* split in centromere. The two chromatids of each chromosome remain together. *Chiasmata* mark the sites of crossing over between chromatids, where exchange of material has occurred. Subsequently chiasmata break off.

Diakinesis

The chromosomes condense further and stain more deeply. This marks the last stage of prophase.

Metaphase I

It begins when the nuclear membrane disappears. Chromosomes move to the equatorial plane of the cell.

Anaphase I

Two members of bivalent (homologous pair) disjoin. At this stage, there is random assortment of maternal and paternal chromosomes. One chromosome from bivalent pair goes to each pole. Chiasmata formation and random assortment of maternal and paternal chromosomes forms physical basis of Mendelian law of independent assortment.

Telophase I

Cytoplasmic division is done in this phase. In 1974, Hulten found that average of 50 chiasmata are seen in spermatocytes.

Meiosis II

It resembles mitosis, but differs from it in two respects. Firstly, there is no DNA replication prior to this. Secondly, the second meiotic division follows meiosis I without interphase. Unlike meiosis I, here centromere splits and two sister chromatids separate to move one to each pole. This results in daughter cells having identical chromosomes. Meiosis II shows similar phases, prophase II, metaphase II, anaphase II and telophase II.

GAMETOGENESIS

Gametogenesis in human beings shows sexual dimorphism. In males, the process is called *spermatogenesis*; in females, it is called *oogenesis.*

Spermatogenesis

It is the process by which spermatozoa are formed. It occurs in seminiferous tubules of the testis after puberty. The wall of the seminiferous tubule is formed by seminiferous epithelium. The latter consists of (*i*) germinal element and (*ii*) supporting element (Sertoli cell). Germinal element consists of a series of developing germ cells. The most peripheral cell population is of spermatogonia. These are stem cells. They are of two types, type A and type B. Type B spermatogonia by their further division form primary spermatocytes. The primary spermatocyte undergoes a prolonged prophase comprising leptotene, zygotene, pachytene, diplotene and diakinesis. It is followed by metaphase, anaphase and telophase (meiosis I). During the anaphase, centromeres do not split. This results in separating members of the homologous pair of chromosomes, each member reaching the opposite pole. This reduces chromosome number to half, i.e. 23 chromosomes. Thus, division of a primary spermatocyte results in the formation of two secondary spermatocytes each with haploid chromosome complement. The secondary spermatocyte soon undergoes second meiotic division (meiosis II) forming two spermatids. Thus, four spermatids are formed from each primary spermatocyte. Spermatids do not undergo division but mature to form spermatozoa. The process of transformation of spermatid into spermatozoon is called *spermiogenesis.* Total time taken from beginning of meiosis to the formation of mature spermatozoa is about 64 days. About 200–300 million sperms are produced per ejaculate.

Oogenesis

This differs from spermatogenesis in certain respects:

1. Oogonia divide to form primary oocytes in prenatal life. No primary oocyte is formed after birth. From prenatal life to puberty, the primary oocytes remain in *suspended prophase.* Meiosis I completes at the time of ovulation. That means some primary oocytes may complete their first meiotic division after 40 years or more. This leaves a chance of meiotic error (nondisjunction) in elderly mothers resulting in higher incidence of numerical/structural aberration in foetus, e.g. Down syndrome.
2. In both meiotic divisions involving oocyte, there is unequal division of cytoplasm. Primary oocyte divides to form one secondary oocyte receiving

most of the cytoplasm and one (first) polar body with hardly any cytoplasm. However, both have haploid chromosome complement.

3. Second meiotic division in oogenesis commences as the female germ cell passes into the uterine tube, but is not completed until after fertilisation. After fertilisation, secondary oocyte completes its second meiotic division to extrude second polar body.

FERTILISATION

It is the process of union of male and female germ cells resulting in formation of zygote. It usually occurs in the ampullary portion of the uterine tube. The process takes about 24 hours to complete. The process of fertilisation involves the following steps:

1. Passage of sperm across corona radiata aided by acrosome reaction releasing hyaluronidase.
2. Passage of sperm across zona pellucida helped by acrosin and zona lysine.
3. Entry of sperm into ooplasm.
4. Occurrence of zona reaction—a physicochemical change rendering zona pellucida impermeable to other sperms (after entry of one).
5. Sperm loses its tail and its head forms male pronucleus.
6. Formation of female pronucleus by extrusion of second polar body.
7. Fusion of male and female pronuclei forming zygote.

Achievements of Fertilisation

1. Reconstitution of species-specific chromosome complement (46 in humans).
2. It allows for biparental inheritance and thus species variation.
3. Initiation of cleavage. Cleavage is a series of mitotic divisions involving zygote and resulting in the formation of smaller cells called *blastomeres.*

So far, we have seen the behaviour of chromosomes during cell division both in mitosis and meiosis. Their behaviour in these two types of divisions is much different. We have also seen how reconstitution of diploid chromosome complement occurs at fertilisation. With this background let us now concentrate upon the structure and analysis of chromosomes.

HUMAN CHROMOSOMES

AN 73.1

We have earlier seen that the human chromosome complement has 46 chromosomes. However, before 1956 the chromosome number was thought to be 48. It was in 1956 that Tjio and Levan with refined cytogenetic techniques convincingly demonstrated that there are 46 chromosomes in human beings. They are in 23 pairs, with 22 pairs of autosomes and a pair of sex chromosomes.

Chromosome Morphology

AN 73.1

Chromosomes are rod-shaped structures, each one consisting of two chromatids. They are held together at the primary constriction, the area that is narrower and in which there is a pale staining region called a centromere. On either side of a centromere, a chromosome has its two arms designated as short arm (p arm) and long arm (q arm). Depending upon the placement of centromere, chromosomes have been classified into four types. They are:

1. **Metacentric chromosomes:** In these, the centromere is almost in the centre and two arms are nearly equal in size.
2. **Submetacentric chromosomes:** In these, centromere is located between midpoint and end of the chromosome.
3. **Acrocentric chromosomes:** They have a centromere very close to one end of the chromosome. Thus, their p arms are very short and q arms are relatively much longer.
4. **Telocentric chromosomes:** These chromosomes have a centromere at one end and have only one arm (Fig. 2.6).

Associated with human acrocentric chromosomes are small round structures (chromatin masses) called *satellites*. They are attached to short arms by narrow stalks called secondary constrictions. Latter contain genes coding for 18S and 28S ribosomal RNA.

Karyotyping

AN 73.2

It is the process by which a karyotype is obtained. In this process, metaphase chromosomes are obtained for analysis and photomicrographed.

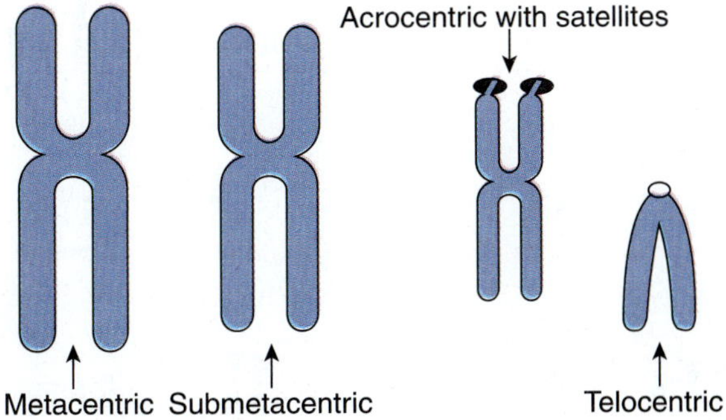

Figure 2.6 Types of metaphase chromosomes. Telocentric chromosomes are not found in human beings.

Photographs of individual chromosomes are then cut and arranged according to standard classification. This is called ideogram or karyotype.

Chromosome classification

AN 73.1

The first attempt towards this was made in 1960 at Denver, Colorado; hence, it is known as "Denver Classification". According to this, the chromosomes are classified under seven groups. They are A-1, 2, 3; B-4, 5; C-6 to 12 and X chromosome; D-13,14,15; E-16,17,18; F-19, 20; G-21, 22 and Y chromosome. The basis of classification includes chromosomal features such as length of chromosomes, placement of centromere and relative lengths of arms (Fig. 2.7). Subsequently in 1971 at the Paris conference, more accurate ways of identifying chromosomes based on various banding patterns were suggested. Today, the Paris nomenclature is accepted all over the world. According to this, both p and q arms consist of regions that are numbered 1, 2 and 3, starting from the centromere. The regions are further subdivided into bands so as to give a precise location, e.g. RBI (Retinoblastoma) locus is situated on chromosome 13. Its precise location is 13q 14, i.e. fourth band on the first region of long arm of chromosome 13. Table 2.1 shows symbols of chromosome nomenclature currently in use.

Chromosome preparation

AN 73.2

Chromosomes can be obtained from somatic cells by culturing them. One can undertake either short-term or long-term culture depending upon the cells used. Under long-term culture, one can use fibroblasts or amniotic fluid

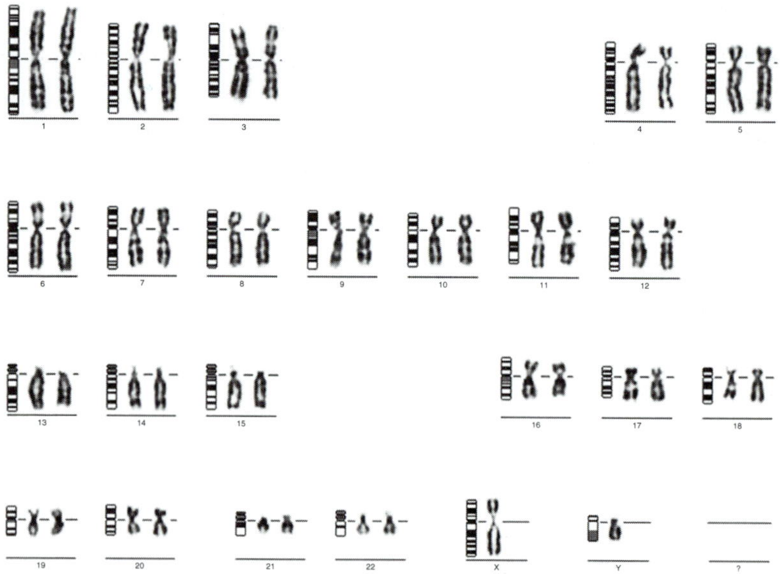

Figure 2.7 Normal male chromosome complement 46, XY arranged according to the standard classification.

Table 2.1 Standard Symbols Used in Describing Karyotype

Symbol	Meaning
p	Short arm of a chromosome
q	Long arm of a chromosome
del	Deletion, e.g. 46, XY, del (5) (p)
dup	Duplication, e.g. 46, XX, dup (13)
i	Isochromosome, e.g. 46, X, i (Xq)
r	Ring chromosome, e.g. 46, XX r (18)
t	Translocation, e.g. 46, XY, t (8;14) (q^{24}; q^{32})
ter	Terminal or end, e.g. pter or qter
inv	Inversion, e.g. 46, XX, inv (9) (p^1 q^{12})
/	Mosaicism, e.g. 46, XX/45, XO

cells. Chromosomes can also be observed directly (without culture) in tissues with high mitotic index like bone marrow or chorion villous samples.

Peripheral blood culture **AN 73.2**

This involves the following steps (Fig. 2.8):

1. **Collection of blood:** Blood is collected from peripheral vein under sterile conditions in heparinised syringe.
2. **Planting:** The blood sample collected is then transferred to culture vials, usually 2–3 vials are set per sample. Each vial contains:
   (a) *Culture medium*: Commonly used media are HAM F10, TC 199, RPMI, etc.

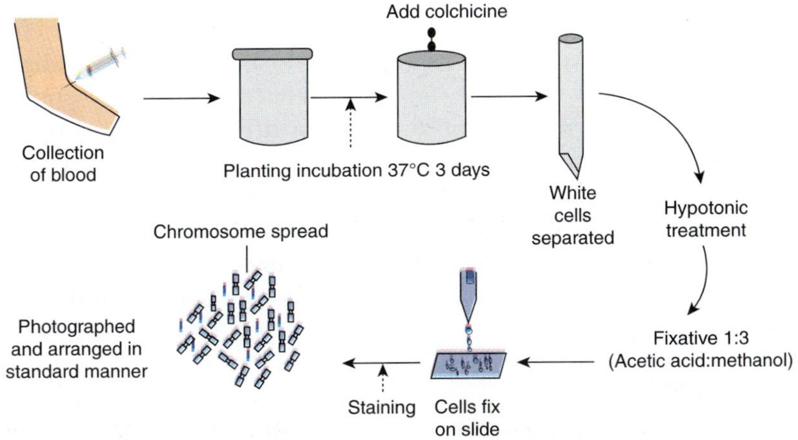

Figure 2.8 Procedure of karyotyping.

(b) Foetal calf serum or alternatively human AB serum may be used. Both (medium and serum) serve to nourish the cultured cells.
(c) *Phytohaemagglutinin*: This is a mitogenic agent. Its addition to a vial is necessary to promote the rate of mitosis of the cultured cells. It is extracted from French bean (*Phaseolus vulgaris*).
(d) *Antibiotics*: Usually penicillin and streptomycin combination serves the purpose but alternatively any other antibiotic can be used. The purpose of antibiotic addition to culture is to prevent bacterial growth.

3. **Incubation:** The culture vials are put in an incubator at 37°C for 3 days. During this period, lymphocytes present in blood undergo mitosis. Cultures should be shaken intermittently during the incubation period.
4. **Harvesting:** Around 70 hours after planting, colchicine is added to the culture vial. This arrests the mitosis at metaphase by preventing formation of spindle tubules. Two hours after addition of colchicine, contents of vial are taken into the centrifuge tube. After centrifugation for 5 minutes, supernatant is discarded and pellet containing cells at the bottom of the tube is treated with hypotonic solution. By this, the cells swell and the chromatids separate. The cells are then re-suspended in fixative. Fixative contains glacial acetic acid and methanol in 1:3 proportions. After three successive changes of fixative, cells suspended in about half millilitre of fixative are dropped from a height on chilled slides.
5. **Staining:** For chromosome analysis, it is essential to stain them with one of the following banding techniques. Among them routinely used is Giemsa banding (G-banding).
   (a) *G-banding*: Slides with chromosome preparation are first treated with a solution of trypsin. Trypsin denatures the chromosome protein. Slides are then stained with Giemsa solution. The chromosomes show dark and light bands that can be observed under a microscope.
   (b) *Q-banding*: In 1970, Casperson demonstrated that chromosomes stained with quinacrine mustard show a specific banding pattern (Q-bands) when observed under fluorescent microscope.
   (c) *R-banding*: In this, the chromosome preparations are preheated in buffer at high temperature and then stained with Giemsa. This gives a banding pattern (R-bands) that is reverse of Q- or G-banding.
   (d) *C-banding*: This method is selectively chosen for staining the centromeric region and other regions with secondary constrictions, e.g. those in chromosome 1, 9, 16 and long arm of Y chromosome (Fig. 2.9).
   (e) *NOR-staining*: In this method, ammoniacal silver is used as stain. It stains nucleolar organising regions, i.e. narrow stalks on some D and G group chromosomes. These regions contain 18S and 28S rRNA genes (Fig. 2.10).

High Resolution Banding **AN 73.2**

In routine metaphase preparation, a total of about 200 bands are seen on the whole chromosome preparation. Instead of this if we take prophase or

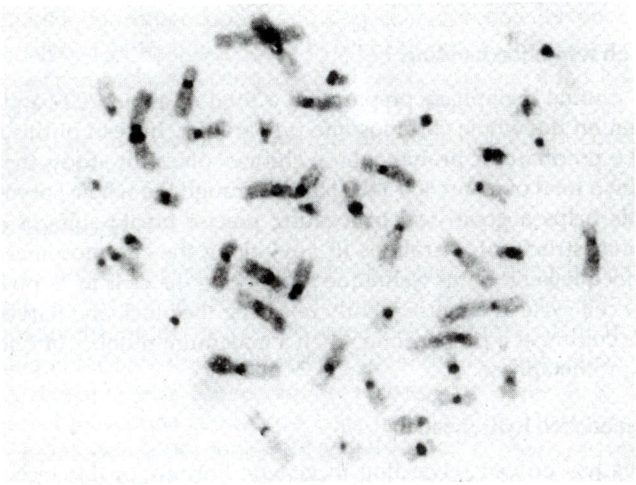

Figure 2.9 C-banding of chromosomes.

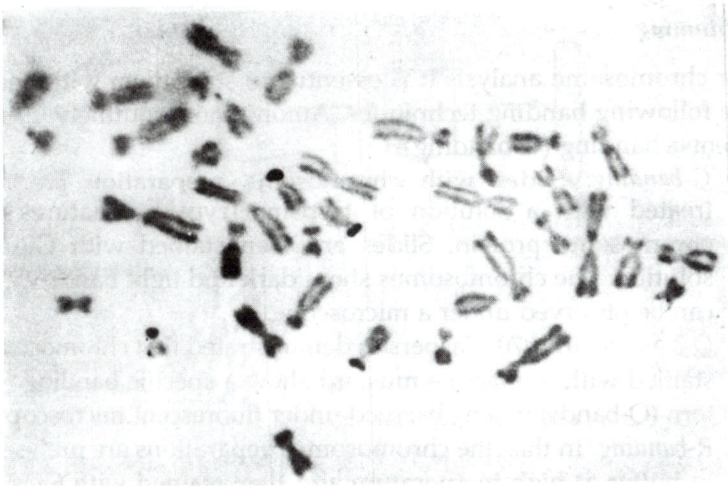

Figure 2.10 NOR staining.

prometaphase chromosomes for study, they exhibit a total of about 800–1400 bands through the whole karyotype. This helps a great deal in locating precise breakpoints or some minor structural alterations (if any) along the chromosomes. The principle behind this technique is—blocking the cells in "S" phase of the cell cycle and subsequently releasing the block and harvesting the culture at a proper time when maximum number of cells are in prometaphase.

Somatic Cell Hybridisation

This has proved rewarding in genetic linkage. In this technique, somatic cells from two different species are fused together under favourable conditions and then cultured (Fig. 2.11). Chromosome complement of the hybrid cell

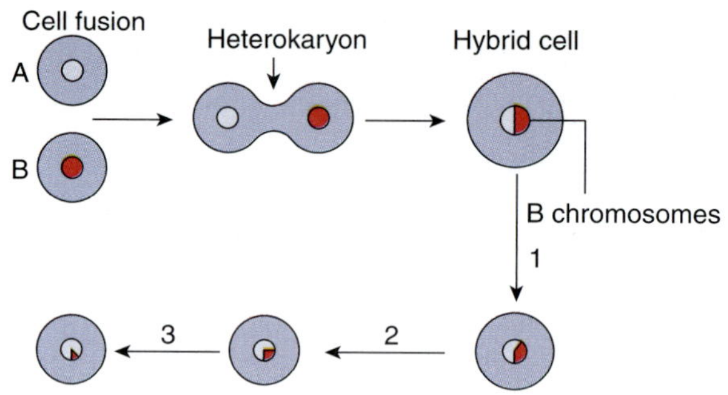

Figure 2.11 Schematic representation of somatic cell hybridisation. Note progressive loss of B chromosomes. 1, 2, 3 represent subsequent nuclear divisions.

is then studied. To cite an example, mouse cells cannot synthesise an enzyme thymidine kinase but human cells can do so. Thus, when mouse cells are fused with human cells, the hybrid cells should possess the capacity to synthesise thymidine kinase and they do synthesise the enzyme. Even after several generations, a hybrid cell retains this capacity. A study of chromosomes from a hybrid cell reveals that it has lost all human chromosomes except number 17. This indicates that the gene coding for enzyme thymidine kinase is located on chromosome 17.

Flow Cytometry

Flow cytometry forms a recent technique that is likely to have a significant impact on present techniques of chromosome analysis. It actually means "fluorescent activated cell sorting" (FACS). In this technique, cells are first ruptured and then stained with a selective DNA dye that is fluorescent in nature. The material is then projected into a fine jet across a laser beam in flow chamber. The laser beam excites chromosomes to fluoresce and the fluorescence can be measured by a detector. The amount of fluorescence depends upon the size of a chromosome. This makes it possible to draw a frequency distribution histogram of the chromosome size with the help of a computer. A distinct advantage of this method is rapid analysis avoiding present day methods that are time consuming. However, the cost of such a unit is prohibitive, thereby limiting its routine use.

Fluorescent In Situ Hybridisation (FISH)

This has revolutionised the concept of chromosome analysis. It is based upon the unique ability of a portion of single-stranded DNA, the probe, to anneal or hybridise with its complementary target DNA sequence located in the genome. This probe is conjugated with a fluorescent label and hence can be visualised under UV light.

Various types of chromosome-specific probes can be used. Some of them are specific for the centromere of chromosome or for a particular portion of a chromosome. Alternatively, probe for the whole chromosome can be prepared. On application to a metaphase spread, this probe hybridises to that chromosome. It is called whole chromosome paint (WCP). This technique can be extremely rewarding for characterising complex chromosome rearrangements such as, deletion, insertion, translocations, ring chromosome, etc.

FISH has an added advantage in that it can be applied at interphase stage, i.e. it can be used to make a rapid diagnosis in conditions like trisomy 21 in interphase nuclei from chorion villous sample (CVS) without culturing the cells. In yet another procedure called "reverse painting", an additional portion of an unidentified chromosome material, such as a small duplication or marker chromosome, is extracted with the help of cell sorter. This is then amplified with polymerase chain reaction (PCR) and used as a probe for hybridisation to a normal metaphase spread. The origin of the unidentified chromosome fragment is then revealed by knowing the chromosome to which it hybridises.

Applications of Karyotyping

AN 73.2

1. **Clinical diagnosis:** Karyotyping helps in reaching a clinical diagnosis. It is especially indicated in patients with congenital malformations involving multiple systems, mental retardation or in cases of ambiguous genitalia.
2. **Gene mapping:** Chromosome analysis has helped in proper localisation of human genes to their specific positions on chromosomes.
3. **Role in cancer:** The detection of Philadelphia chromosome in patients with chronic myelogenous leukaemia (CML) alters prognosis. The formation of Philadelphia chromosome involves translocation between long arms of chromosomes 22 and 9. The Philadelphia-positive CML cases have a longer survival than those without Philadelphia chromosome.
4. **Repeated foetal loss:** On chromosome analysis, the couple may reveal a chromosomal defect in any one of the partner. Chromosomal aberrations account for a sizeable number of spontaneous abortions in the first trimester of pregnancy.
5. **Prenatal diagnosis:** Chromosome analysis of chorion villous samples and amniotic cells may reveal a chromosome abnormality in a foetus warranting medical termination of pregnancy.

SEX CHROMATIN

In 1949 Barr and Bertram, while studying cat neurons, found that some of these cells show a chromatin mass in their nuclei. This was observed only in females but not in males. Subsequently, it was labelled as sex chromatin or Barr body. The cells that contain this are called chromatin-positive and others are called chromatin-negative cells. Barr body can be found in many cell types but can be conveniently examined in buccal mucosa.

Procedure of Examining Barr Body

It is relatively simple. Scrapping from the inner side of the cheek is taken on a slide and smeared evenly. Subsequently, it is fixed in alcohol and stained

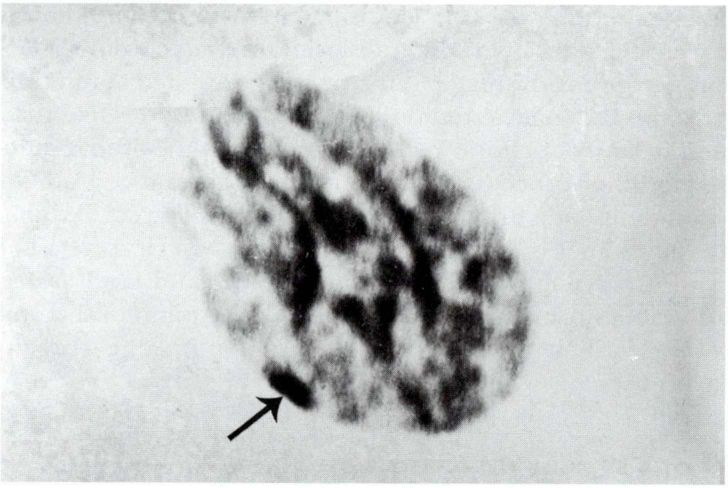

Figure 2.12 Barr body.

with thionin. It is then mounted in neutral medium and observed under a microscope (Fig. 2.12).

In the past, buccal smear (sex chromatin study) was used as a diagnostic tool for disorders of sexual development. However, now it has been replaced by karyotyping. Number of Barr bodies in a cell will depend upon the number of X chromosomes in the cell, i.e. number of Barr bodies = number of X chromosomes – 1. For example, in an individual with 47, XXX complement, there are 3 X chromosomes. Therefore, the number of Barr bodies is 3 – 1 = 2. A Turner syndrome patient having 45, XO complement has only one X chromosome. Therefore, the number of Barr bodies is 1 – 1 = 0, i.e. no Barr body.

Barr body represents one of two X chromosomes of a female cell. This remains condensed and is in inactive state throughout interphase. Its replication is also late as compared to its homologue.

Lyon's Hypothesis

AN 73.3

Dr Mary F. Lyon, in 1962, stated in her hypothesis about the inactivation of X chromosome. It was observed that at prophase, one X chromosome is late replicating and heteropyknotic, i.e. this X chromosome differs from other chromosomes in respect to state of condensation and staining. Of the two X chromosomes only one is active in cellular metabolism while the other (inactive one) forms sex chromatin. In males, there is only one X chromosome, which is active and hence they do not show Barr body.

Mary F. Lyon

(Used from PLoS Genetics. Attributed to: 2010 Jane Gitschier.)

Lyon's hypothesis states that:

1. In female somatic cells, only one X chromosome is active. The second is inactive, condensed and appears in the form of sex chromatin in interphase.
2. Inactivation occurs early in embryonic life.
3. Inactivation is random but fixed. The inactive X can be maternal or paternal (X^m or X^P) in different cells of the same individual. However, once the decision as to which X will be inactivated is made in the cell, then all the clonal descendants of that cell will follow the decision. Sex chromatin is detected in blastocyst at 9–12 days. First it is detectable in syncytiotrophoblast, then chorionic mesoderm followed by yolk sac. In embryo proper, it is detected after the 18th day.

Mechanism of Inactivation of X Chromosome

AN 73.3

It involves DNA methylation; specifically, cytosine undergoes change forming 5-methyl cytosine at certain sites in DNA. This accounts for altered gene activity. DNA methylation is said to be responsible for inactivation of genes; the exact mode of which is yet not clear.

Inactivation Centre

AN 73.3

It is believed to be in the proximal part of the long arm of X chromosome (X_q) around which the Barr body condenses. The evidence attesting this fact is as follows:

1. An abnormal X lacking proximal part of X_q has not been observed to form a Barr body.
2. An abnormal X with duplication of this part of X chromosome forms a bipartite Barr body.

There appears to be a couple of exceptions to "random inactivation". In X chromosome abnormality such as deletion, ring chromosome or isochromosome, it is the abnormal X that forms a Barr body. Accordingly, the size of the Barr body may be larger or smaller than the normal one. Secondly, in translocations involving X chromosome and an autosome, it is the intact X that becomes inactive to form a Barr body.

Genetic Significance of X Inactivation

AN 73.3

It is threefold:

1. Dosage compensation
2. Variability of expression
3. Mosaicism

Dosage compensation

The X chromosome inactivation explains why the X-linked gene product is equivalent in both sexes in spite of two X chromosomes in female and only one in male.

Variability of expression

As inactivation is random, female heterozygotes for X-linked genes present a considerable phenotypic variation. Variation in expression of X-linked disorders can be ranging from completely normal to full expression of the defect. A carrier who exhibits an X-linked trait is called manifesting heterozygote, e.g. colour blindness, haemophilia, Duchenne muscular dystrophy.

Mosaicism

AN 75.2

Mosaic is an individual with at least two cell lines with different genotypes but derived from a single zygote.

Females are mosaics in respect to X chromosome. They possess two cell populations, one cell line with one X chromosome active and the other with an alternative X active. Davidson et al. (1963) demonstrated mosaicism by cloning cultured fibroblasts from a woman heterozygous for two different G6PD alleles.

There are some X-linked genes that do not get inactivated. They are Xg locus for Xg blood group and STS locus for steroid sulphatase. These loci are located on the distal end of the short arm of X chromosome (Xp). It is suggested that they escape inactivation through a mechanism of X–Y pairing during meiosis.

Origin of X-Inactivation

AN 73.3

It is thought that initially X and Y were homologous through most of their length. Subsequently, part of Y became involved in testicular development and part of X became concerned with ovarian development. Afterwards, part of Y that was not involved in testicular development got translocated to X chromosome. This led to a duplication of genes and long arm of X chromosome. To prevent the double dose effect, it became necessary to inactivate that part of X chromosome in somatic cells.

Summary

1. **Cell cycle** consists of *mitosis–interphase–mitosis*.
   Interphase consists of G1, S and G2 phases.
   Mitosis consists of:
   i) *Prophase*: Chromosomes condense and centriole divides.
   ii) *Metaphase*: Formation of metaphase plate and spindle formation.
   iii) *Anaphase*: Chromatids disjoin with vertical split of centromere, cytoplasmic division starts with a furrow at equator.
   iv) *Telophase*.
2. **Sister Chromatid Exchange (SCE):** It involves crossing over between sister chromatids of single chromosome.
3. **Meiosis** consists of:
   Meiosis I (reduction division).
   - It is much prolonged and at the end of it number of chromosomes is reduced from diploid (46) to haploid (23).

Summary—cont'd

- It has prolonged prophase I comprising proleptotene, leptotene, zygotene, pachytene, diplotene and diakinesis stages. Prophase I is followed by metaphase I, anaphase I (without centromeric split) and telophase I.

Meiosis II—It is like mitosis except that there is no interphase and no replication of DNA.

4. **Oogenesis** differs from spermatogenesis in the following respects:
   - All primary oocytes are formed before birth and remain suspended at prophase.
   - There is unequal cytokinesis (division of cytoplasm).
   - Meiosis II of oogenesis occurs only when sperm enters ooplasm.
5. **Fertilisation:** Process of union of mature male and female germ cells, resulting in formation of zygote.
   This results in (*i*) reconstruction of species-specific chromosome complement—46 in humans; (*ii*) species variation through biparental inheritance; (*iii*) sex determination; and (*iv*) initiation of cleavage.
6. **Chromosome Morphology:** *Types*—metacentric, submetacentric, acrocentric and telocentric (depending on place of centromere).
   Karyotyping: It is the process of obtaining chromosomes. *Steps*: (*i*) Collection of blood; (*ii*) planting; (*iii*) incubation; (*iv*) harvesting; and (*v*) staining and then observing/studying under microscope.
7. **High Resolution Banding (HRB):** Helps in detection of structural alterations.
8. **Somatic Cell Hybridisation:** Helps in the study of genetic linkage and localisation of gene/s on chromosome/s.
9. **Flow Cytometry:** It involves "fluorescent activated cell sorting (FACS)". It allows much faster chromosome analysis.
10. **Applications of Chromosome Analysis:** (*i*) confirming clinical diagnosis; (*ii*) gene mapping; (*iii*) prognosis in cases of CML; (*iv*) in repeated foetal loss; and (*v*) prenatal diagnosis.
11. **Barr Body:** Number of Barr bodies in a cell = Number of X chromosome – 1; e.g., for a cell with 46, XX it is 2 – 1 = 1.
    Lyon's hypothesis:
    In female somatic cell, only one "X" chromosome is active while the other is inactive, condensed to form Barr body.
    Inactivation occurs in early embryonic life.
    Inactivation is random but fixed.
    Mechanism—DNA methylation.
    Inactivation centre—Proximal part of long arm of X chromosome.
    Significance of X inactivation: (*i*) Dosage compensation; (*ii*) variability of expression; and (*iii*) mosaicism.

QUESTION YOURSELF*

1. Euchromatin represents:
   a. Extended pale staining portion of chromosomes
   b. Coiled dark staining portion of chromosomes
   c. Both coiled and extended portions of chromosomes
   d. None of the above
2. How is the word chromosome derived?

3. What is cell cycle?
4. In which phase of cell cycle do most of the cells in the body exist and function?
5. What are the characteristics of interphase?
6. Name two salient features of metaphase.
7. What is the striking feature of anaphase in mitosis?
8. What is the role of colchicine in karyotyping?
9. What is somatic recombination?
10. What is sister chromatid exchange?
11. What is the striking feature of anaphase of meiosis I?
12. What is the difference between meiosis II and mitosis?
13. Which one of the following is true about the primary oocytes?
    a. All primary oocytes are formed in prenatal life.
    b. All primary oocytes are formed at puberty.
    c. All primary oocytes are formed after menarche.
    d. Primary oocytes are formed continuously from puberty until menopause.
14. In which phase of meiosis I, the primary oocytes remain suspended?
    a. Prophase
    b. Metaphase
    c. Anaphase
    d. Telophase
15. Which one of following chromosome complement is normally present in the daughter cells resulting from meiosis I?
    a. Polypoid
    b. Aneuploid
    c. Diploid
    d. Haploid
16. Second meiotic division in oogenesis usually completes when the female germ cell is in ________
    a. Ovary
    b. Ampulla of uterine tube
    c. Uterine cavity
    d. Cervical canal
17. What is fertilisation?
18. "Human chromosome complement has 46 chromosomes" was established in the year ________
    a. 1936
    b. 1952
    c. 1956
    d. 1969
19. Name the four types of chromosomes depending upon the placement of centromere.
20. Which one of the following banding techniques is routinely used in chromosome analysis?
    a. C-banding
    b. G-banding
    c. Q-banding
    d. R-banding

21. After Q-banding, the chromosome spread is observed under________
    a. Binocular research microscope
    b. Polarised microscope
    c. Fluorescent microscope
    d. Electron microscope
22. Which chromosomes show secondary constrictions?
23. What is the advantage of high resolution banding (HRB)?
24. What is somatic cell hybridisation?
25. What is flow cytometry?
26. What is FISH?
27. What are the applications/indications for karyotyping?
28. What is Barr body?
29. How many Barr bodies are seen in a cell?
30. What is Lyon's hypothesis?
31. How is inactivation of X chromosome achieved?
32. Where is the inactivation centre?

---

*See pages 257–258 for Answers.

3 Molecular Genetics

COMPETENCY AN 75.4 Describe the genetic basis of mutation.

LEARNING OBJECTIVES

At the end of this chapter the students should be able to understand

- Structure of DNA, its types and transcription and translation
- Structure and types of RNA and its role
- Gene structure and its function
- Mutation, recombinant DNA, polymerase chain reaction

KEY WORDS

DNA, RNA, Gene, Recombinant DNA, PCR, DNA sequencing, Gene chip

In this chapter, an attempt has been made to answer few basic questions. For example, what is a gene? How does gene determine a character, like colour of eyes and skin, height, etc.? What are the events involved in the production of protein, the final product of gene? To find out the answer to these questions, let us first equip ourselves with some basic information.

Each human cell has a nucleus that stores genetic information (Fig. 3.1); it is surrounded by cytoplasm containing various organelles such as mitochondria, ribosomes, endoplasmic reticulum, etc. Let us look into the pages of history for evidence to support our statement that nucleus stores genetic information. Friedrich Miescher in 1869 conducted some experiments. He made chemical analysis of nuclei obtained from pus cells. He detected a substance with a high phosphorus content. It was called nuclein. Later on, it was called nucleic acid considering its acidic properties. In 1928, Griffith performed an experiment on pneumococci. Morphologically, there are two forms of this bacteria—rough (R) and smooth (S) form. Griffith showed that if S form of pneumococci are killed and the remains are mixed with living R form, then some of the R bacteria are transformed into S form. Now the question is—what brings about this transformation of R bacteria into S form? Logically, it is only the material from dead S bacteria that can do so. This was confirmed by further experiments in 1944 by Avery et al. They purified extracts from dead S bacteria. They further showed that the transformation from R to S form could be brought about by using nucleic acid solution. This

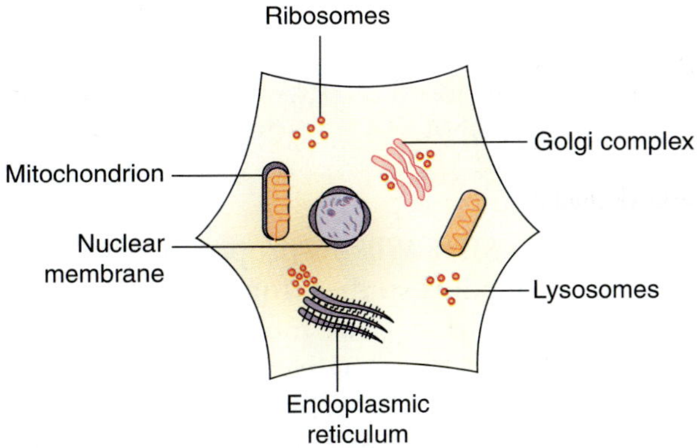

Figure 3.1 Schematic representation of eukaryotic cell.

confirmed that the genetic information regarding this transformation was passed through nucleic acid. This was further supported by experiments on bacteriophage (Fig. 3.2) in 1952 by Hershey and Chase. They labelled the outer protein coat of phage with radioactive sulphur and inner nuclei with radioactive phosphorus. It should be noted that the protein coat has practically no phosphorus and nucleic acid has no sulphur. These labelled phages were allowed to infect bacteria. On analysing these infected bacteria, it was discovered that the radioactive phosphorus of nucleic acid had entered the bacteria. The radioactive sulphur from the protein coat was not found. The infected bacterial cells could, however, produce complete bacteriophage particles with the protein coat. This means that the information for synthesis of the protein coat was passed through the nucleic acid (with a radioactive phosphorus label).

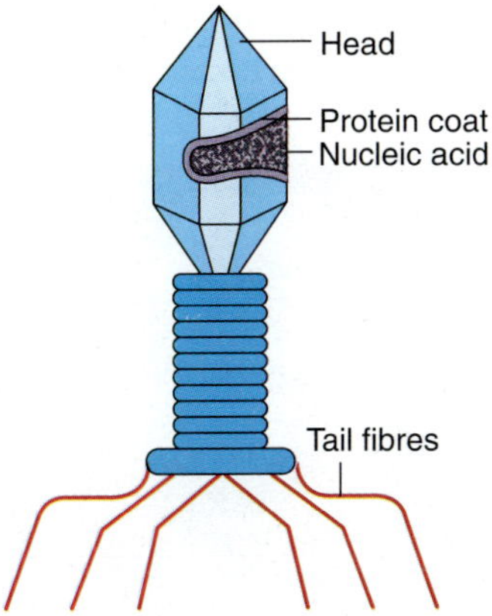

Figure 3.2 A diagrammatic representation of bacteriophage.

STRUCTURE OF NUCLEIC ACID

It consists of long chains of molecules called nucleotides. Each nucleotide molecule in turn consists of a nitrogenous base, a sugar moiety and phosphorus molecule. The nitrogenous bases are of two types—purine and pyrimidine. The purine bases are adenine and guanine; the pyrimidine bases are thymine, cytosine and uracil. There are two types of nucleic acids—deoxyribonucleic acid (DNA) and ribonucleic acid (RNA). DNA contains sugar called deoxyribose, and RNA contains sugar called ribose. DNA is mainly found in chromosomes (exception mitochondrial DNA) and RNA is chiefly found in nucleolus and in cytoplasm. Among pyrimidine bases, DNA has thymine and RNA has uracil; the rest of the nitrogenous bases are common in both.

DNA

We realise that genes are composed of DNA, and so it is essential at this stage to consider the structure of DNA.

Structure: Watson–Crick model

JD Watson, FHC Crick and MHF Wilkins proposed a structure of DNA molecule based on X-ray diffraction studies, for which they were awarded the Nobel Prize. They suggested that DNA molecule consists of two chains of nucleotides in the form of double helix. Each chain has a backbone of sugar–phosphate molecule (Fig. 3.3). The two chains are held together by hydrogen bonds between nitrogenous bases. The DNA chains have polarity due to orientation of sugar–phosphate molecules. The 5′ end of the chain is terminated by a 5′ carbon atom of a sugar molecule and the 3′ end of the chain is terminated by a 3′ carbon atom. In the two strands of DNA double helix, the 5′ end of one strand is opposite the 3′ end of the other. In short, the strands are said to be anti-parallel.

FHC Crick
(*Attributed to:* Gene Forum.)

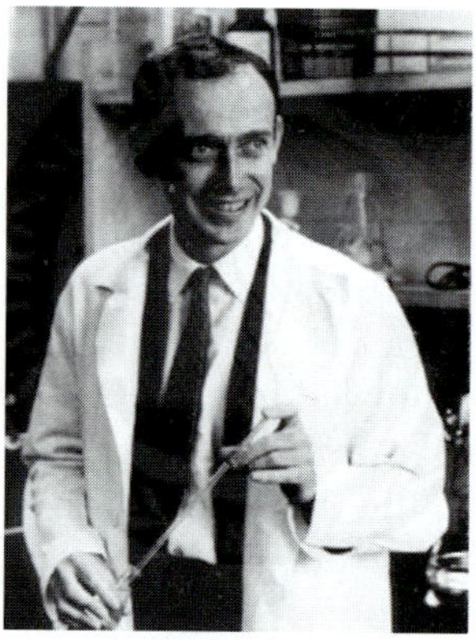

James Watson
(*Attributed to:* Marc Lieberman.)

MHF Wilkins
(*Used from:* Leyo. *Attributed to:* Website Der National Institutes of Health.)

The bases in DNA molecule pair with specificity. Adenine always pairs with thymine and cytosine with guanine. The two strands of DNA molecule separate at a nuclear division, and each strand then builds up its complement. This is called replication. The process of DNA replication commences simultaneously at multiple sites along the length of DNA strand and then progresses in both directions.

Types of DNA

There are three main types of DNA:

1. Unique sequences
2. Satellite DNA
3. Interspersed repetitive DNA sequences

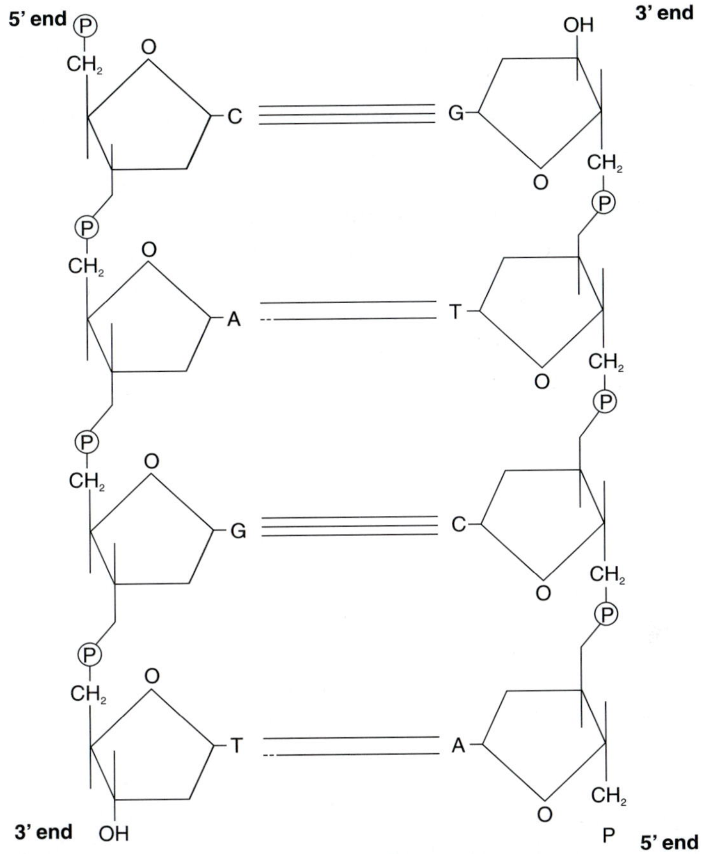

Figure 3.3 Diagrammatic representation of sugar–phosphate backbone and nucleotide pairing in DNA duplex. A = adenine, C = cytosine, G = guanine, T = thymine and P = phosphate.

Unique Sequences

It has been estimated that there are about 50–100 thousand genes that code for specific proteins. They form unique sequences and account for about 5% of DNA of the human genome. Out of the rest 95%, about 75% consists of again unique or single copy DNA sequences, precise function of which is yet unknown.

Satellite DNA

About 10%–15% of human genomic DNA comprises of short tandem repeat (STR) DNA sequences that code for ribosomal and transfer RNAs (tRNAs). These sequences are clustered in the heterochromatic regions of chromosomes 1, 9, 16 and long arm of Y chromosomes. This DNA separates out on density gradient centrifugation as a shoulder or "satellite" to the main peak of DNA and hence referred as satellite DNA.

Interspersed Repetitive DNA Sequences

This accounts for about 10%–15% of the human genomic DNA. It is made up of two main classes of repetitive DNA sequences, which are interspersed throughout the genome. Out of these, one class is made up of short interspersed repetitive elements (SINEs) that form about 5% of human genome. Each sequence is about 300 bp (base pairs) and is in nearly 300,000 copies. They are called "Alu repeats" because they contain an Alu I restriction enzyme recognition site. "Alu" is a bacterial restriction enzyme.

Another class of interspersed repetitive DNA elements is made up of LINEs, i.e. long interspersed repetitive sequences or LI family. They are about 6000 bp in length and occur in nearly 100,000 copies. The function of these sequences is not clear. However, both Alu and LI family sequences (i.e. SINEs and LINEs) have been implicated as cause of mutations in inherited human diseases.

"Selfish DNA", it has been referred so because it preserves itself as a result of selection within the genome but appears to have little or virtually no function and does not seem to contribute to the phenotype.

Mitochondrial DNA

Hundreds of mitochondria in the cell possess their own DNA. In a zygote, they are from the cytoplasm of an oocyte (maternal in origin). The mitochondrial DNA is circular and is often designated as mtDNA. It codes for two types of ribosomal RNAs, protein sub-units of some of the enzymes e.g. cytochrome B and cytochrome oxidase, involved in oxidative phosphorylation. It also codes for numerous tRNAs. Recently, mitochondrially mediated diseases are being recognised, e.g. Leber optic atrophy. It has been shown to be due to a defect in detoxification of cyanide owing to mutations in mitochondrial DNA. This involves gene for electron transport protein NADH-coenzyme Q oxidoreductase. Since mitochondria have an important role in cell metabolism, mitochondrial DNA mutations are observed in central nervous system, heart and skeletal muscle.

Pedigree in Fig. 3.4 shows mitochondrial inheritance/cytoplasmic inheritance. The disorder affects both males and females but is transmitted only through females.

Chromosome

The width of chromosome is much larger than the diameter of DNA molecule. With this consideration, there must be tight coiling of DNA helices. The *Solenoid model* of the chromosome structure was proposed by Finch and Klug. It is believed that a total length of DNA in chromosomes, if extended, runs into metres while a total length of human chromosome complement is

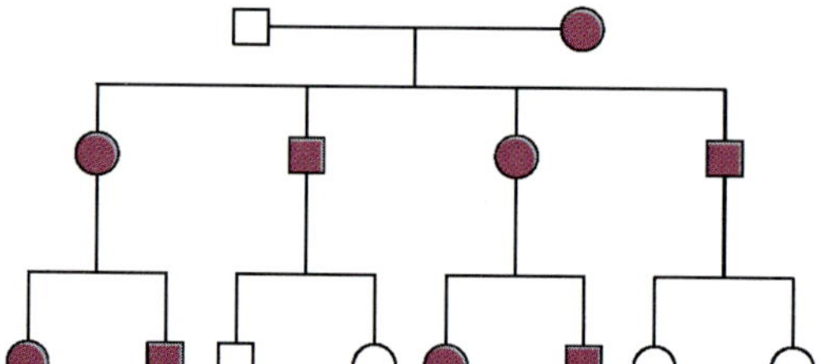

Figure 3.4 Pedigree showing mitochondrial inheritance.

about half a millimetre. This means that there may be several orders of DNA coiling in chromosomes. They are as under:

1. Primary coiling of DNA double helix.
2. Secondary coiling around histone beads, i.e. nucleosomes.
3. Nucleosomes undergo tertiary coiling to form chromatin fibres.
4. Chromatin fibres in turn form loops.
5. These loops further coil tightly to form a chromosome, which we observe under a microscope (Fig. 3.5).

Code

The term "code" is defined as the sum of available signals by means of which news or information can be formulated and transmitted to the recipient.

Triplet code

In 1961, Nierenberg and Matthaei took the first step to find key to the genetic code. They added RNA containing only the nitrogenous base uracil (U) to a mixture of amino acids, enzymes and ribosomes; the result was synthesis of a simple protein containing only the amino acid phenylalanine, even though many other amino acids were available in the mixture. With this, the investigators arrived at the conclusion that triplet UUU codes for the amino acid phenylalanine.

The basic function of a gene is to direct synthesis of protein. There are 20 different amino acids in protein. DNA molecule stores genetic information in the form of a triplet code, which is a sequence of three bases that code for one amino acid. As there are four bases—A, T, G and C—there can be 4^3 = 64 such combinations. The sequence of these three bases is also called genetic code or codon. For some amino acids, there is more than one triplet code. In such cases, codes are sometimes called degenerate. What happens if there is mutation of gene? Then the resulting code may be read as "nonsense" and no amino acid shall be coded. Alternatively, the code may be read as "missense"; in this case, a different amino acid is substituted resulting in an abnormal protein. Two of the codons, UAA and UAG, are called *Ochre*

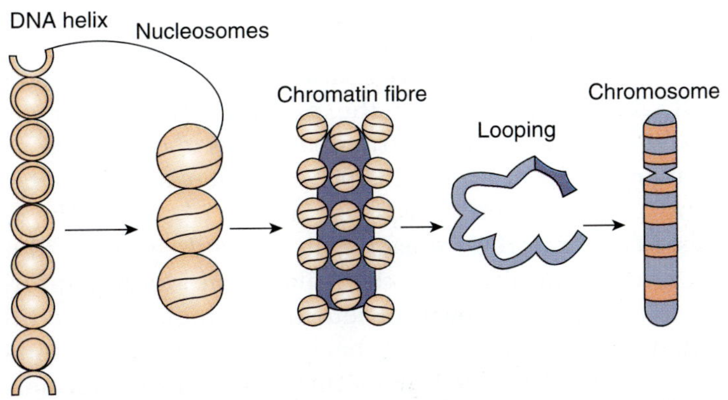

Figure 3.5 Schematic representation of "Solenoid model" of DNA coiling.

and *Amber*, respectively. They are found in micro-organisms. They do not code for any amino acid.

Ribonucleic Acid (RNA)

It differs from DNA basically in three respects:

1. It has sugar ribose in place of deoxyribose of DNA.
2. Of the four bases, three are common in DNA and RNA. They are adenine, cytosine and guanine. The fourth base in DNA is thymine; in RNA, it is uracil.
3. RNA molecule is usually single stranded, while DNA has two strands.

Types of RNA

RNAs are of three types:

1. mRNA—messenger RNA
2. tRNA—transfer RNA
3. rRNA—ribosomal RNA

Messenger RNA and Heterogeneous Nuclear RNA (hnRNA)

The primary product of transcription is a heterogeneous nuclear RNA (hnRNA) molecule. It has been shown that hnRNA contains coding and non-coding sequences. Coding sequences are called "exons", while non-coding sequences are called "introns". Non-coding sequences are removed by excision, i.e. cut and removed. Coding sequences (exons) are then spliced together to form a mature form of RNA called mRNA. The mRNA undergoes couple of modifications to become mature. Firstly, it acquires a methylated cap at the 5′ end. Secondly, at the 3′ end, a long sequence "poly-A tail" gets attached. The methylated cap consists of a cluster of methyl groups. It possibly protects mRNA against degradation at the 5′ end. Poly-A tail consists of a long sequence of about 200 adenine residues. It helps in transporting mRNA from the nucleus to the ribosomes in the cytoplasm. This is attested by the fact that mRNA, which lacks poly-A tail, does not reach cytoplasm.

Transfer RNA (tRNA)

The message derived from DNA is mRNA, but this mRNA message cannot be directly read by amino acids. Crick and Hoagland first thought of an intermediate RNA molecule serving the purpose. This molecule is called tRNA. A tRNA molecule is single stranded. It has about 73–93 nucleotides and a molecular weight of about 25,000. A particular tRNA attaches to its specific amino acid and then transports it to the ribosome, the site of protein synthesis. In other words, for each amino acid the tRNA is different. For some amino acids, more than one tRNA has been identified; for example, more than five tRNAs are known for leucine. A single stranded tRNA molecule is folded upon itself forming hairpins. Many tRNAs show unusual bases like inosine, pseudouracil or methylated forms of normal bases. All tRNAs possess a terminal sequence of CCA at the 3′ end. This is called *amino acid arm*. Apart from this, tRNA possesses three loops:

1. *TC loop* containing pseudouracil residue
2. *DHU loop* containing dihydrouridine
3. *Anticodon loop* containing specific triplet of bases complementary to codon of mRNA (Fig. 3.6)

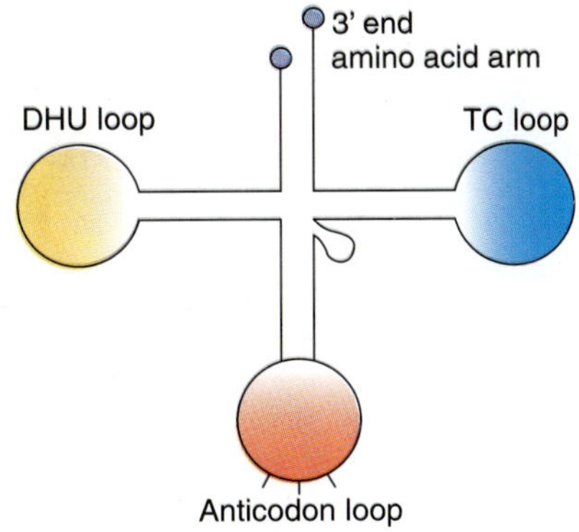

Figure 3.6 Clover leaf model of tRNA molecule.

Ribosomal RNA (rRNA)

Ribosomes are tiny particles of ribonucleoprotein in cytoplasm. In eukaryotes, they occur in two sizes—60S in mitochondria and 80S in cytoplasm. Ribosomes consist of two unequal subunits. Ribosome, in cytoplasm, may lie free or attached to endoplasmic reticulum. They are also present in the nucleus in the form of nucleolus. Ribosome consists of rRNA and ribosomal protein. Electron micrographs of protein-synthesising cells show ribosomes associated in a string-like manner; these are called polyribosomes. The latter are formed by attachment of ribosomes to a single molecule of mRNA. Ribosomal RNA present in subunits of ribosome is in a highly folded form. Approximately, 70% of rRNA is double stranded. Apart from being structural component of ribosomes, rRNA is involved in protein synthesis. The 3′ end of 16S rRNA has a sequence complementary to mRNA ribosome binding site. The 5S rRNA in larger subunit of ribosome possesses a sequence complementary to the TC loop sequence of tRNA. This permits binding of tRNA to ribosomes.

TRANSCRIPTION AND TRANSLATION

Transcription (Fig. 3.7)

It is a process whereby information is transmitted from DNA to the mRNA. This occurs in the following manner:

1. Two strands of DNA double helix separate.
2. Against the single strand of DNA (opened up), there is synthesis of mRNA molecule. This occurs with complementary base pairing—cytosine pairs with guanine, thymine with adenine, but adenine pairs with uracil (since RNA does not have thymine but has uracil).
3. The mRNA then migrates from the nucleus to the cytoplasm.

Translation (Fig. 3.8)

It is a process of translating information from mRNA into protein synthesis. For this, the mRNA formed within the nucleus comes out into the cytoplasm:

1. Here it gets associated with ribosomes. These are the sites of protein synthesis. A group of ribosomes associated with an mRNA molecule is called polyribosome.

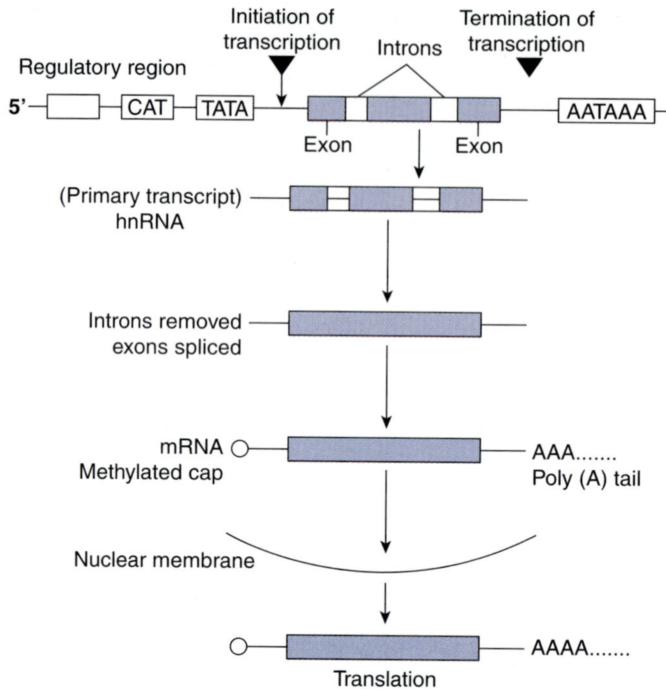

Figure 3.7 Diagrammatic representation of eukaryotic transcription unit and the way it operates.

Animation 1: Transcription. (*Available online at* www.medenact.com)

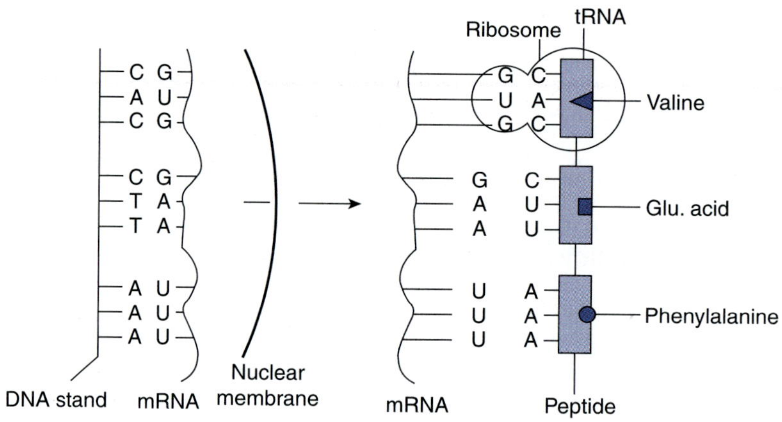

Figure 3.8 Translation of genetic information into protein synthesis.

Animation 2: Translation. (*Available online at* www.medenact.com)

2. mRNA serves here as a template and hence is also called template RNA.
3. Amino acid to be incorporated in protein gets activated by ATP.
4. tRNA present in the cytoplasm receives activated amino acid at one end.
5. tRNA at its other end has three bases, complementary to the mRNA bases.
6. Thus mRNA triplet through tRNA triplet picks up a required amino acid.
7. The ribosome with tRNA + amino acid moves along mRNA molecule. This links the amino acid in a polypeptide chain. After being formed, the

polypeptide chain is released and the ribosome moves to a new point along the mRNA molecule.

GENE

Until 1977, gene was thought to be a segment of DNA molecule possessing a code for amino acid sequence of a polypeptide chain. This model, however, appears to be inadequate to elucidate the precise mechanism of gene operation.

There are about 50,000–100,000 DNA sequences that code for RNA or protein products in humans. These are called structural genes. They range from 1000 to 200,000 bp size. On analysis, a structural gene (Fig. 3.9) reveals coding sequences called *exons*, which are interrupted by non-coding sequences called *introns*. Introns are initially transcribed but are not represented in mature mRNA or in the final protein product. The gene also possesses extensive *flanking regions*. They are important in regulation and the "start" and "stop" signals.

Let us now consider the structure of human globin gene to understand gene structure (Fig. 3.9). Human globin gene presents three exons (coding sequences) and two introns (non-coding sequences). Upstream (5′ end) from the gene there is a flanking region containing two specific sequences. They play a significant role in regulation. They are as follows:

1. *CAT box.* It is located 70 bp upstream from the beginning of first coding sequence (exon).
2. *TATA box.* It is located between the CAT box and the site of initiation of transcription.

Downstream (3′ end), there is a second flanking region, containing a sequence AATAAA. This sequence appears to be a signal for addition of poly-A tail to the end of mRNA strand.

It would be interesting at this stage to know about what happened in 1970. Khorana and his colleagues successfully synthesised a gene de novo. They

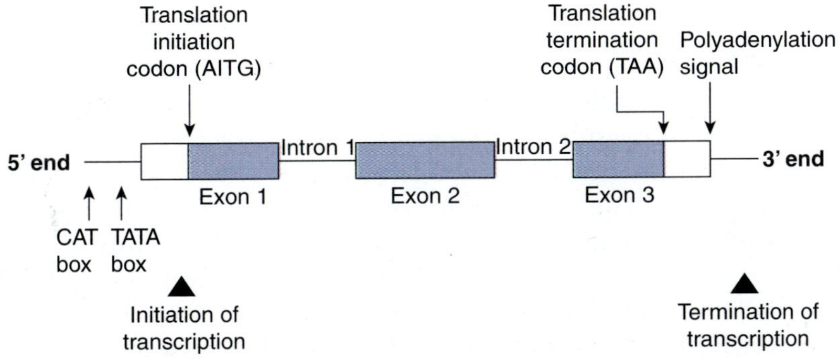

Figure 3.9 Schematic representation of the structural gene in humans.

assembled 77 bp of gene that codes for production of alanine tRNA in yeast. The artificial gene so synthesised, though structurally correct was non-functional because it lacked "start" (initiator) and "stop" (termination) signals. However, with some more research, they succeeded in synthesising fully functional artificial gene.

Generalisation of Gene Structure

1. In genes, coding sequences are split by intervening sequences.
2. The exon–intron patterns of split genes appear to be strikingly conserved during evolution, e.g. α and β globin genes have arisen by duplication of a primitive precursor about 500 million years ago. Each of them has two introns at precisely the same locations.
3. Within exons, alterations in sequence occur slowly, approximately at the rate of 10^{-9} substitutions per codon per generation through natural selection.

The concept of "split gene" was introduced by Richard J. Roberts and Phillip A. Sharp; they were awarded Nobel Prize in 1993 for the same.

Phillip A. Sharp
(*Attributed to:* Paloma Baytelman.)

Richard J. Roberts
(*Attributed to:* Chemical Heritage Foundation.)

HTF Islands

A restriction enzyme called HpaII contains one or more CpG dinucleotides in the nucleotide recognition sequence necessary for DNA cleavage. Methylation of cytosine in this dinucleotide pair prevents cleavage of DNA by restriction enzymes. The CpG dinucleotides are non-randomly distributed in the genome. The clusters of under methylated CpG dinucleotides are found near transcription initiation sites at 5′ end of many genes. These are called methylation free HpaII tiny fragments or HTF islands. However, all genes do not have HTF islands and not all the HTF islands are associated with genes.

Jumping Genes

Also called "transposons or transposable genetic elements"; these are regions of DNA that can jump to and fro within single chromosome or an adjacent one. The discovery of jumping gene goes to the credit of Barbara McClintock, an American geneticist during her maize plant study. Later on, similar moving genetic elements were also discovered in bacteria. In bacteria, these are involved in rapid spread of antibiotic resistance genes. Transposons are supposed to be ubiquitous. These mobile nucleic acid elements are of great significance in search for vectors. The latter are carrier molecules that help in transport of the desired genes into the host cell.

Pseudogenes

These are sequences that show a striking similarity with functional genes, but are not transcribed. Probable reason being that their regulatory regions have been altered by mutation. They are thought to represent vestigial remains of the gene that was functional at some stage of evolution. Some pseudogenes do not have introns; they are called "processed pseudogenes". They represent cDNA (complementary DNA) sequence synthesised on mRNA template. It is then reinserted into the genome.

Control of Gene Action in Prokaryotes "Operon"

In 1961, Jacob and Monod worked extensively on *Escherichia coli* (prokaryote) at Pasteur Institute. They postulated that apart from structural genes, there is another class called *control* genes. These are regulator genes and operator genes. They regulate the expression (action) of structural genes. In short, they govern the amount of (protein) product produced by gene (Fig. 3.10).

A unit of functional gene consists of an operon that in turn comprises operator gene and structural gene. The operator gene controls the action of the structural gene. The regulator gene produces a substance called "repressor", which inhibits the operator gene. Therefore, when the regulator gene is in functional state, no proteins are produced by the structural gene. It shall

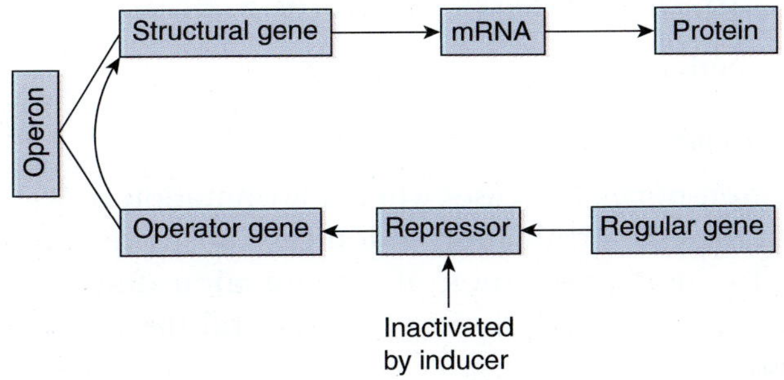

Figure 3.10 Concept of operon by Jacob and Monod (1961).

function only when the regulator gene is switched off. This is achieved by the inactivation of the repressor by another substance. It is called *inducer.*

In human beings, an example of operon mutation is possibly sucrose intolerance. However, gene activity appears to be controlled more likely by the flanking regions, i.e. CAT and TATA box regions as read earlier in this chapter.

MUTATION

AN 75.4

It is defined as any change in sequence of genomic DNA. In the normal course, DNA replication is highly precise; however, any error that involves this process is copied in subsequent replications. Mutation occurs by any of the following three mechanisms: substitution, deletion and insertion.

Single base substitution or "point mutation"

AN 75.4

In this, there is a single base change in DNA sequence. This alters the triplet code and causes replacement of amino acid. Fortunately, since the code is degenerate (more than one triplet codes for amino acid), all the single base substitutions shall not alter the amino acid sequence in the final protein product. For example, leucine and arginine are specified by six different codons, and so a single base substitution in this will alter only one of the triplets but the remaining five remain normal. About 20%–25% of the single base mutations belong to this type. In the rest of the 75%, there is a change in amino acid and thus an alteration in the final protein product. Such a protein may not have any biological activity. If the gene is responsible for a particular enzyme, the enzyme may be synthesised in lesser quantity. Rarely will a mutation cause increased synthesis of an enzyme, e.g. Hektoen variant of G6PD.

Deletion or insertion

AN 75.4

The deletion or insertion of a base leads to an alteration in the reading frame of DNA strand and thus the amino acid sequence. This is called "Frame shift". Usually, such massive alteration does not result in active proteins.

Mutations and Their Effects on Gene Product

AN 75.4

They are described as follows.

Chain termination mutations

DNA transcription normally ceases when a termination codon is reached. A mutation that creates a termination codon can cause premature cessation of transcription. In another situation, if the mutation destroys a termination codon then it allows transcription to continue till the next termination codon is reached.

Splice mutations

These are mutations that affect a normal mechanism by which introns are excised and exons are spliced together during the formation of mRNA. It leads to complete failure of synthesis of the gene product.

Mutations involving regulatory sequences

Mutations that involve CAT box and TATA box regions upstream of the structural gene can cause reduced transcription of the sequence. One can say that mutations may lead to a state where the protein product is synthesised, but this protein is functionally inactive. This happens in haemophilia A patients. They have Factor VIII of coagulation, but not a functionally active one.

Gene Mapping

It can be broadly divided into two. The first is "chromosome mapping", i.e. assigning a particular gene or a DNA sequence to a specific chromosome. The second is a finer piece of information. It includes physical relationships to flanking DNA sequence polymorphisms and the detailed structure of the gene, i.e. "DNA mapping".

Chromosome mapping

Chromosome mapping may be accomplished by somatic cell hybridisation or in situ hybridisation (see details under Chapter 2).

DNA mapping

For this purpose, numerous techniques are available, such as pulsed field gel electrophoresis (PFGE), chromosome linking/jumping, YAC contigs and so on.

Pulsed Field Gel Electrophoresis (PFGE)

The routine agarose gel electrophoresis can resolve DNA fragments of about 20,000 bp length. Using PFGE, one can separate DNA fragments of 2,000,000 bp in size. The digestion of DNA with restriction enzymes such as Not 1 and Pvu 1 results in producing larger DNA fragments. This is because these enzymes have 6–8 bp nucleotide recognition sequences, which occur less frequently in DNA. This allows construction of maps of relatively larger stretches of DNA that are not amenable to resolution by routine restriction mapping methods.

Chromosome Jumping

It is a technique used in the physical mapping of genomes. Circular DNA is produced by digesting DNA fragments with restriction enzyme in presence of a plasmid, cut with the same restriction enzyme.

The circular DNA thus obtained is cut again with second restriction enzyme that does not cleave in plasmid sequence. This plasmid sequence acts as a tag and allows cloning of the ends of the original DNA fragments that along with complementary libraries can be used to map markers, which are many kilo bases away. This is called chromosome jumping/linking.

YAC Contigs

Yeast artificial chromosome (YAC) has allowed cloning of large segments of genomic DNA. This is essential in mapping genes, and their flanking regions have 2,000,000–3,000,000 bp length, for example, genes of cystic fibrosis, neurofibromatosis and dystrophin, etc. Cloning of large genomic DNA fragments into YACs has been used in long-range physical mapping for gene cloning by production of overlapping YAC clones or contigs.

Gene Cloning

If a gene in question is required in large quantities, the genetic engineer inserts the gene he wishes to clone into carrier molecule capable of passing through a membrane called "vector". The vector, like a legendary Trojan horse, penetrates the membrane of the host cell, carrying with it the desired gene to be cloned. If the host cell accepts the vector and the genetic command of the smuggled gene, it then starts to copy this gene with its own synthesis apparatus. Under suitable conditions, the host cell begins to produce protein molecules that correspond to the genetic information provided by the cloned gene. Thus, the cloned gene is "expressed" or translated into proteins.

Gene Bank

It is the collection of artificial, recombined (recombinant) DNA molecules that taken together possess the complete genetic information for a given organism. These DNA fragments are preserved in the form of plasmids, either in bacterial cultures or in bacteriophages. "Gene bank" is constructed by cutting open the hereditary material of a particular cell with the help of restriction enzymes. The individual pieces of DNA are then incorporated into the bacterial plasmids and introduced into the host bacteria. As long as the bacterial culture thrives and grows (by reproduction), the plasmid DNA and the foreign DNA inserted into it also increases. In case of need, it can be isolated from the bacterial culture. This can be useful in maintaining the species that are on the verge of being extinct.

Indian National Gene Bank

It provides facility for long-term ex situ conservation of base collections. It is set up by National Bureau of Plant Genetic Resources (NBPGR). It has four components, which are as follows:

1. **Seed repository:** The seeds are dehydrated at 15°C and 15% RH to around 5% moisture content, sealed in laminated aluminium foils under vacuum and preserved at –20°C.
2. **Tissue culture repository:** Healthy (disease free) tissue cultures of recalcitrant seeds (not able to withstand dehydration and low temperature) and

plant species, developed from meristem explants are maintained using growth-limiting media and by suitably sub-culturing them.

3. **Cryopreservation facility:** Samples of "synthetic seeds", embryos, gametes can be preserved by this technique. This is possible either by immersion in liquid nitrogen (–196°C) or by keeping in its vapour phase around –150°C. This is however according to the protocols using computerised freezing–thawing rates and also chemical cryoprotectants.
4. **Clonal repository:** Germplasm collections of vegetatively propagated crops like ginger, turmeric, sweet potato and banana are being maintained.

India is one of the biodiversity rich countries. The plant wealth in India and in Indian Gene Bank can be imagined that it held 132,619 dehydrated, sealed seed samples till 1994. Accessions still under multiplication and characterisation in NBPGR network recorded that time were 120,226, and more than that 188,263 were still under multiplication and upgrading at Indian PGR system.

Recombinant DNA

Among recent advances in the field of genetics, recombinant DNA occupies a prominent position.

Definition

It is an artificially synthesised DNA that is constructed by insertion of foreign DNA into DNA of an appropriate organism so that foreign DNA is replicated along with the host DNA.

Restriction enzyme

Restriction endonucleases are enzymes that can cleave DNA at specific sites. They were discovered by Hamilton Smith and his associates in 1970. Today more than 200 of them are known.

Vectors

They are used to carry foreign DNA fragments. They are as follows: plasmids, phages, cosmids and yeast artificial chromosomes.

Plasmids

Plasmid is a circular extrachromosomal element in bacteria. It can replicate independently. Plasmids vary in size. One of the plasmids from *E. coli* known as pBR 322 is 4362 bp in length. Plasmids have an advantage as a vector in that they have a limited number of unique restriction sites and can also carry resistance to particular antibiotics. This character is used to identify recombinant clones.

Phages (Bacteriophages)

These are viruses that infect bacteria and multiply within bacteria. Subsequently, they cause lysis of bacterial cell. This releases a phage progeny that then infects another bacteria. Phages have an advantage over plasmids that much larger fragments of DNA can be cloned in them. Plasmids can be used as vectors up to 8 kb (1 kb = 1000 bp). Phages can be useful for fragments up to 15 kb; for larger fragments of 35–45 kb, cosmids are used.

Cosmids

These are plasmids in which the maximum DNA has been removed to permit largest possible insert for cloning but still have DNA essential for in vitro packaging into phage particle. In short, they are phage–plasmid hybrids capable of carrying relatively large DNA inserts.

Yeast Artificial Chromosomes

YACs consist of plasmid possessing DNA sequences essential for (*i*) centromere formation and (*ii*) telomere formation and other DNA sequences known as autonomous replication sequences. YACs can incorporate DNA fragments up to 1000 kb (kilobase) in size. The choice of vector used in cloning depends on number of factors such as the restriction enzyme used and the size of DNA fragment to be inserted.

Recombinant DNA procedure

It involves five steps (Fig. 3.11):

1. DNA is cut into fragments by restriction endonuclease enzyme.
2. Incorporation of these fragments in a suitable vector using DNA ligase.
3. Transformation of host organism, e.g. *E. coli* by recombinant vector, i.e. the recombinant vector (plasmid) is reinserted into the host cell by exposing the host (bacterial) cell to calcium salts. This turns the cell permeable to plasmid.

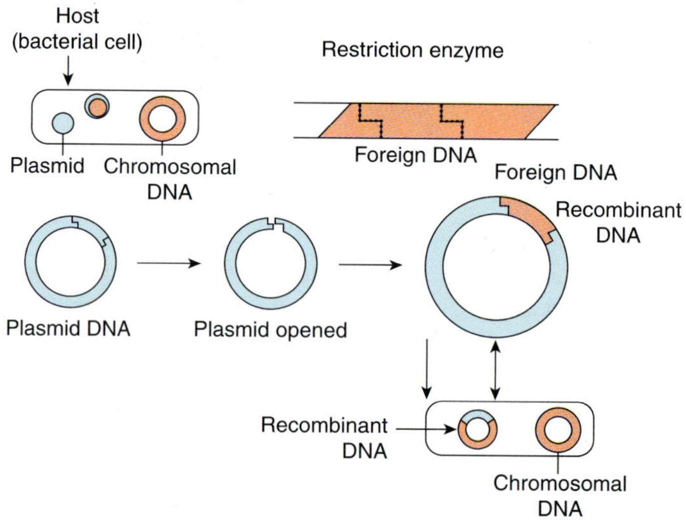

Figure 3.11 Constructing recombinant plasmid and transformation of host cell.

4. The host cells containing vectors are put in a culture medium to produce clones, i.e. obtain multiple copies of foreign DNA fragments incorporated in recombinant vector.
5. Selection of clones possessing proper DNA fragment.

Applications of recombinant DNA technology

1. It gives us a rational approach to the understanding of molecular basis of numerous diseases, e.g. sickle cell disease, familial hypercholesterolaemia, thalassaemias, cystic fibrosis and Huntington chorea, etc.
2. Using this technology, human proteins can be produced for therapeutic purposes, e.g. insulin and growth hormone.
3. Production of proteins for vaccination, e.g. hepatitis B.
4. Proteins can be produced for diagnostic tests, e.g. AIDS test.
5. Gene therapy for sickle cell disease or thalassaemias or other diseases.

DNA probe

The process of making recombinant DNA involves the insertion of a large number of random restriction fragments into plasmids. This obviously makes it important to be able to recognise a particular gene. It is possible, if mRNA of the desired type is available. With the help of an enzyme, reverse transcriptase, a cDNA can be transcribed from a known RNA. The enzyme, reverse transcriptase, is obtained from RNA tumour viruses. The final product of the transcription is a radioactive cDNA molecule, also referred to as DNA probe. The probe hybridises to the specific gene or its mRNA under suitable conditions.

POLYMERASE CHAIN REACTION

The PCR was discovered by Kary Mullis and developed by Saiki and others in 1985. It has revolutionised both the diagnostic as well as therapeutic ability of physicians. It has become a powerful tool in the field of molecular biology. Before we turn to PCR, let us acquire little background knowledge of DNA synthesis.

The DNA synthesis requires a template (to be copied), a primer, which is extended as a newly synthesised DNA, dNTPs, Mg and an enzyme—DNA polymerase. The DNA is synthesised by polymerase action, i.e. by adding one nucleotide at a time and the chain extends.

Primer: It is a short stretch of nucleotides having complementary sequences. It attaches to a long template molecule at specific region. The 3′ end of the primer gets extended by addition of an appropriate nucleotide, i.e. T (thymine) if the template has A (adenine), G (guanine) if the template has C (cytosine). The 5′ end of the primer remains fixed. The new DNA is synthesised only in 5′–3′ direction.

PCR: Polymerase chain reaction is an in vitro method of synthesis of nucleic acids, wherein, a specific DNA segment is amplified rapidly without concomitant replication of the rest of the DNA molecule. Thus in PCR, a limited region of a DNA molecule is amplified. Basic requirement for this assay is nucleotide sequence information at two ends of the part to be amplified. Then a pair of oligonucleotide primers (about 20–30 nucleotide long) are then synthesised. One of them is complementary to upper/one strand of DNA at one end and the other primer is complementary to the other end of the lower/other strand of the DNA to be amplified.

The reaction tube [containing test DNA, primer pair, dNTPs, Mg, buffer and thermostable DNA polymerase (Taq polymerase) enzyme] is put in a thermal cycler that is programmed to carry out amplification.

Standard Reaction

The reaction is conducted in 0.5 mL Eppendorf tubes; the following components are added and the reaction volume is made up to 100 µL.

1. Template (100 ng–1 µg DNA)	10 µL
2. Buffer (containing KCl, Tris-HCl, $MgCl_2$ gelatine)	10 µL
3. Primer 1 (20–50 pmol)	10 µL
4. Primer 2 (20–50 pmol)	10 µL
5. dNTP mix (200 µm of each dNTP)	10 µL
6. Taq polymerase (2.5 units)	0.5 µL

Two drops of mineral oil are then added to overlay the reaction mixture to avoid evaporation.

Procedure

The amplification procedure involves three steps: (*1*) denaturation, (*2*) annealing and (*3*) extension (Fig. 3.12).

1. **Denaturation:** Since the test DNA is double stranded, it has to be converted into single stranded one. This is achieved by heating the DNA at 92–95°C for about 30–60 seconds. This breaks the hydrogen bonds between the complementary bases, thus separating the two strands. This does not damage the DNA in any way.
2. **Annealing:** The cycler is then set to a lower temperature i.e. 55–65°C for about 30–60 seconds. Since the primer concentration is very high in the tube, the primer will anneal with A=T and C=G base pairing, to the test DNA at the specified locations (as per primer). The high concentration of the primers prevents reannealing of the original (test) DNA strands. The annealing temperature is critical and depends upon the composition of the DNA sequence and the length of primers.
3. **Primer extension (DNA synthesis):** In this step, the annealed primers are extended. The DNA chain, as learnt earlier, grows in 5′–3′ direction. The nucleotides are added one at a time. The nucleotide sequence of new DNA is decided by the sequence of template DNA. Since the primers are designed to extend in opposite directions, the intervening DNA (between the two primers) is synthesised.

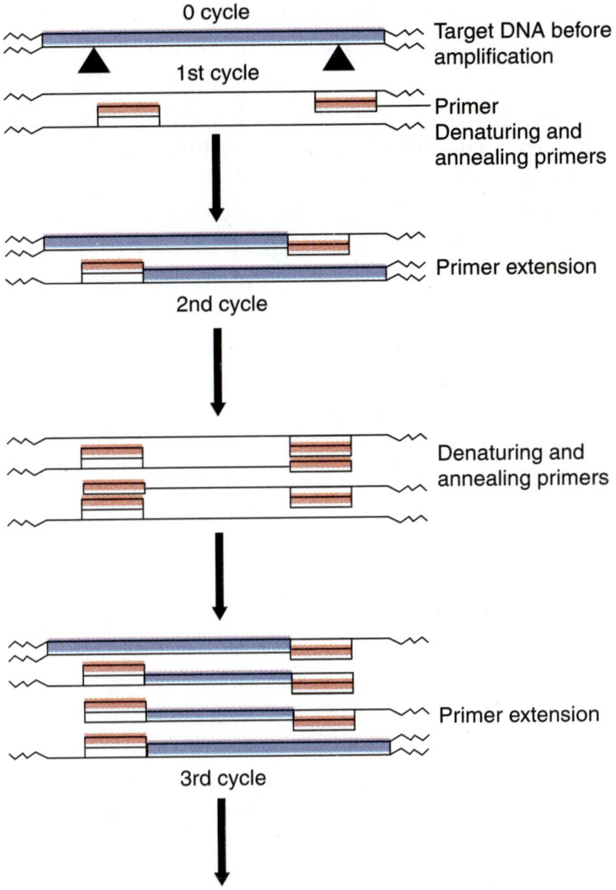

Figure 3.12 Schematic representation of polymerase chain reaction (PCR).

Usually this step is carried out at 72°C, because most of the commercially available thermostable DNA polymerases have optimum activity at this temperature. The incubation time at this temperature depends upon how farther the primers are placed along the DNA molecule.

These three steps make up one cycle in PCR reaction. At the end of this, the original DNA molecule under consideration is quantitatively doubled. The thermal cycler is programmed to repeat the cycle 25–30 times. With each cycle, the desired DNA (between two primers) is doubled. Therefore, it is possible to selectively amplify any given DNA several million times in few hours.

Analysis of PCR Product

This is one of the most important steps. Sometimes, the primer/s bind to non-specific locations if the annealing temperature is lowered, which favours annealing of mismatched bases. Analysis is accomplished as under:

- **Estimation of Size:** The PCR product is subjected to the agarose gel electrophoresis and its size is estimated by running known DNA size markers.

The newly synthesised DNA should be of the size between the extreme ends of primers. Mostly it is sufficient to conclude between the positive and negative results.

- **Nested PCR:** The PCR product can be reamplified using a new primer pair that is located within the ends of the first primer pair. This PCR product is shorter than the first PCR product.
- **Hybridisation:** A short oligonucleotide may be synthesised, which is complementary to one of the DNA strands of the PCR product. Such oligonucleotide is radiolabelled, and the hybridisation signal can then be obtained on autoradiography.
- **Restriction Enzyme Mapping:** The PCR product may be subjected to restriction digestion in order to confirm the identity of the product. This also gives us information regarding RE site gained or lost due to mutation/s.
- **Cloning and Sequencing:** Alternatively, the PCR product can be cloned in an appropriate vector and its authenticity can be verified by nucleotide sequencing. The sequencing of PCR product can be done even without cloning.

Advantages of PCR

1. The technique is rapid as well as sensitive.
2. A small quantity of template DNA (5–10 ng) is also sufficient.
3. A highly purified DNA sample is not essential.
4. Number of samples can be used e.g. peripheral blood, bone chips, single sperm, hair follicle and even paraffin embedded tissues.

Problems of PCR

False Positive Reaction: Any contamination in the sample can be amplified giving false positive signal. This could be from the previous reaction or from the exogenous source.

False Negative Reaction: This may be due to very low yield or the absence of the specific product. The conditions of PCR amplification may be altered to overcome this. If essential, even the sequence of the primers can be changed.

Applications of PCR

It can be used in:

1. Diagnosis
2. Therapeutics
3. Criminology, etc.

Amplification Refractory Mutation System (ARMS) PCR

It is a special type of PCR in which the 3′ end base of the primer is mutant specific. The sequence of the primer is selected in such a way that its 3′ end base matches with the mutant base under question. Thus two primers, a mutant and a normal, are used.

SOUTHERN BLOT TECHNIQUE

It is used in DNA analysis. The technique was evolved in 1975 by Edwin Southern at Edinburgh. The steps involved in this technique are as follows (Fig. 3.13):

1. DNA is cleaved by restriction enzymes.
2. DNA fragments are separated by agarose gel electrophoresis. Small fragments move faster than the large ones.
3. DNA is then denatured with alkali. This makes the DNA single stranded.
4. Denatured DNA (single stranded) is then transferred on nitrocellulose filter by blotting.
5. Now to identify and localise a particular fragment on the filter, a radioactive labelled DNA probe p^{32} is used.
6. Probe is allowed to hybridise with DNA fragments in "Southern blot" and subsequently autoradiographed (Fig. 3.13).

NORTHERN BLOTTING

In this technique, mRNA is isolated and run on an electrophoretic gel and is transferred to a filter. This is called Northern blotting. Hybridisation of the

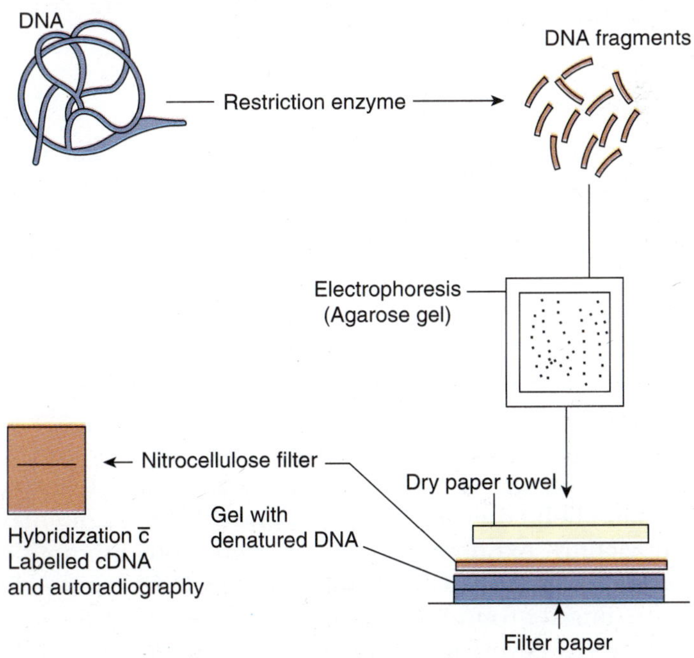

Figure 3.13 The Southern blot technique.

Northern blot with a radiolabelled probe allows determination of the size as well as the quantity of the mRNA transcript. Some single gene disorders in which no mutation has been identified in exons (coding sequences), an alteration in the size of mRNA transcript possibly goes in favour of mutation in non-coding regions (introns) of the gene, like at splice junction of an intron–exon border.

DNA FINGERPRINTING TECHNOLOGY

A British geneticist Dr Alec Jeffreys pioneered DNA fingerprinting technology in 1985. It has now become a routine method for crime investigations in all parts of the world. It is used for identification of the criminals. Using DNA from the suspect and from the site of crime, if the two match perfectly, one can conclusively say that the suspect under consideration is the criminal. However, there are two prime questions before an ordinary person accepts the statement. The first, how reliable is the technology and secondly, how fool proof are its conclusions. To answer these questions, one can recapitulate some well-known facts.

People differ from each other only at about 1% nucleotide locations. In human genome, there are over 3 billion nucleotides. Even 1% of 3.3 billion amounts to 33 million such locations. Considering that at each of these locations, there are four alternative nucleotides that may occur. Further, one knows that there are two DNA strands that make up DNA of a person. Making all the permutations and combinations, one can safely assert that the complete DNA sequencing of any two individuals on the earth cannot be identical unless they are identical twins.

Hypervariable Regions of the Genome

In our genome, there are hundreds of specific regions. At each of these, the nucleotide arrangement differs from individual to individual. Literally, thousands of possible types could be observed. With this technology, one can reliably type individuals at several such regions either sequentially or simultaneously. Combining these observations, one gets almost an error-free DNA profile to identify an individual.

Introduction of DNA typing in courts rests upon similar genetic principles. It has high level of discrimination among individuals and can be performed with many source materials, e.g. bones, skin, hair follicles, saliva, epithelial swabs, etc. Therefore, it is superior to traditional blood typing. Stability of DNA molecules allows genetic typing even from fossilised or decomposed materials. This has helped in establishing identity from fragmented material from the war victims, exhumed bodies or even from mummified bodies. Identification of members of Czar family through DNA typing of excavated bodies from the burial ground is a glaring example of this technology.

Sir Alec Jeffrey's discovery was instrumental in determining the nature of gene loci that can be subjected to such typing offering higher degree of discrimination. In our genome, short stretch of nucleotides are repeated one after another several times. The stretch can be as short as a single base pair or can be long enough to be of hundreds of nucleotides. What differs between people is the **number of repeats of such sequences**. The hypervariability (extreme variation) comes from how many nucleotides are within a single repeat and how many times this unit is repeated close to each other.

In the 1970s, an approach was developed according to which it is possible to shear a long DNA molecule so that each end of such repeat regions have signatures, which indicate the exact/specific location of the cut DNA molecule. It is called digestion, with specific restriction enzymes that cut DNA molecule into small number of nucleotides (4–6 bp) found in a specific order. Since 99% of the nucleotides in the genome are invariable between people, it is possible to ascertain which restriction enzyme would cut the DNA at unique positions in all the people. The DNA molecule (genomic) when digested with specific restriction enzyme, the digestion product consists of thousands of pieces bearing signature of the specific recognition sequence of nucleotides, characteristic of the enzyme used for digestion. The pieces thus obtained are sorted out by size in an electric field through the technique called electrophoresis. Synthetically produced short nucleotide sequences/probes unique to specific regions of genome are now available. Using them for hybridisation with DNA fragments of specific regions, one can isolate regions of specific interest. This is Southern blot technique, named after Dr Edwin Southern. It detects variation (polymorphism) of restriction enzyme digested DNA lengths and is called Southern blot RFLP analysis.

The pairing task (hybridisation) can be accomplished by using either **multilocus probe** that will identify simultaneously repeat regions from a large number of genomic regions or **single locus probe**, identifying only a single region. Probe is radioactively labelled, so that it can be observed on autoradiography. Each fragment thus derived has a repeat region of variable number of tandem repeats (VNTRs). They are flanked on each side by unique DNA sequences of constant lengths. The size variations of the fragments are indicators of how many repeat units such regions contain. This technique needs large DNA strands so that sheared DNA molecules are intact. However, in heavily decomposed body or an old and badly preserved one, it is difficult to obtain large intact DNA fragment. Solution being simple, DNA extracted from even a single cell may be subjected to PCR amplification. It needs artificially produced DNA templates or PCR primers. PCR product is then subjected to Southern blot RFLP analysis, to know number of repeat units and the sequence composition of such repeated regions.

In one case in Finland, the victim's body was decomposed and the DNA sample from the anal swab was not suitable for RFLP analysis. The geneticist who helped the investigation in this case recovered small quantity of DNA from saliva residues on a cigarette butt and rim of the wine glass. The DNA samples were subjected to PCR amplification. The PCR product was typed for several STR markers to show match with the blood from the suspect.

As against single locus typing with RFLP/PCR-based analysis, if the same analysis is done with multilocus probes, the DNA profiles on an autoradiograph appear like bar codes on tags of items in departmental store. Matching of individuals can then be done by the extent to which the bands are shared.

The perfect matching of all fragment sizes indicates identical origin of DNA. A partial band sharing implies their biological relationship. Since even a single multilocus probe usually reveals many bands, the chance of establishing identity from perfectly matching parallels are similar to that of the conventional fingerprinting. It is for this reason that Sir Alec Jeffrey in his paper in 1985 coined "DNA fingerprinting".

Applications

1. DNA fingerprinting has been widely used in forensic science.
2. Number of genetic disorders are now recognised by this technology before their clinical onset.
3. Tissue matching, essential before transplantation, can be done with this technique.
4. The entire Human Genome Project relies upon these techniques to find gene locations.
5. Gene therapy, the future modality of treatment, also rests upon the fidelity and reliability of these techniques.

DNA SEQUENCING

It is knowing the exact nucleotide sequence. In this, the target DNA is amplified before sequencing. There are two ways by which this can be done.

Manual Sequencing—Sanger Method

This was practiced before the advent of automatic sequencers. For manual sequencing, 2,3-dideoxy chain termination method is commonly used.

DNA to be studied is converted into single stranded DNA templates, the aliquots of which are added to four tubes, each containing polymerase primers, and the four deoxynucleotides, i.e. dATP, dCTP, dTTP and dGTP, with radioactive label in some of them. One more type, dideoxy nucleotide triphosphate (ddNTP) is added to each tube. DNA replication starts in each tube, but stops when ddNTP is incorporated into the newly synthesised strand. This is because ddNTP lacks hydroxyl group at 3′ carbon position. For example, in a tube containing ddGTP, there will be various DNA fragments at different positions of guanine in the chain.

This is followed by polyacrylamide gel electrophoresis (PAGE) of the amplified products from each tube. It is then autoradiographed. The band pattern is read from the bottom of the gel to top, to determine the sequence synthesised DNA which is complementary to that of original DNA strand.

Automated Sequencing

This is done with the help of automated sequencing machine. In this, the nucleotides are labelled with fluorochromes (fluorescent dyes). This is a rapid and accurate method of sequencing. Another advantage of this is that different fluorochromes are used to label the four base specific reactions; hence, all of them can be loaded in a single lane. The output is detected by computerised laser detection system. It is in the form of intensity profiles for each of the differently coloured fluorochromes.

Automated sequencing has remarkably changed the speed of sequencing. It can be used for detection of mutations leading to single gene disorders.

MICROARRAY TECHNOLOGY

It is used to study expression of many genes at the same time on a single slide, called DNA chip. Any physiological or pathological state involves an interaction between many genes. Some genes are upgraded while some of them are downregulated. Study of expression of these multiple genes is essential but complex. However, it can be achieved through the DNA microarray technology.

In this, a solid surface such as glass is used for microarray (DNA chip) and detection is done by fluorescent labelling. With the robotic technique, probes are placed on a slide. About 1 cm^2 area on the slide can accommodate thousands of different nucleotides. These are attached to the chip in a specific structured manner called "microarray".

mRNA is extracted from the sample to be tasted, it is then transformed into cDNA and labelled with fluorescent dye. Simultaneously mRNA from normal cells is also extracted and transformed into cDNA. This cDNA obtained from the normal cells is labelled with different fluorochromes.

Both the cDNAs are then mixed, and then hybridised to probes in the microarray. With lasers, the fluorochromes are excited. The pattern of excitation is different at each spot depending upon the ratio of cDNA segment in the test and normal cDNA. The digital images of the colour pattern of the microarray are captured and analysed using a computer.

Application of this gene expression study is of great significance in pharmacogenetics to develop drug molecules and also in cancer research. It is helpful in the diagnosis of single gene disorders associated with multiple mutations. One of the first applications of high-powered "gene chip" technology is in an important psychiatric syndrome, in which scientists reported the discovery of genes that may prove key to understanding schizophrenia.

People, not populations, will be treated with tailor-made drugs that suit their genetic makeup; gene chips will identify who is at more risk of disease, so they can have more frequent checkups; similar chips will distinguish one type of cancer from another, so the best treatment is chosen; gene transplants will be used to correct mutations that cause metabolic disorders.

CRISPR-Cas9

Background

CRISPR stands for

- C—Clustered
- R—Regularly
- I—Interspaced
- S—Short
- P—Palindromic
- R—Repeats

CRISPR refers to a family of DNA sequences encountered in prokaryotic genomes bacteria or archaea, for example. These sequences are derived from bacteriophages which have previously infected the bacteria. These are helpful in detection and subsequent destruction of DNA from similar bacteriophages in the following infections. Thus, CRISPR sequences subserve a role in antiviral (antiphage) defence mechanism in prokaryotes.

Cas9: It is a CRISPR-associated protein. It is an enzyme which uses CRISPR sequences and forms a basis of technology called CRISPR-Cas9, which helps in editing genes in an organism. This naturally has wide-ranging applications in biomedical research, as well as in the health care system, for management of many diseases. It forms an efficient biotechnology tool in genetic disorders which were earlier unconquered. CRISPR-Cas forms immune system in prokaryotes to combat against foreign elements such as the one in phages or plasmids. It forms a sort of acquired immunity.

CRISPR-Associated Systems

It was observed that the prokaryote repeat cluster was accompanied by Cas genes. Four Cas genes have been identified. The Cas proteins exhibiting helicase and nuclease motifs suggest that CRISPR loci have dynamic structures. CRISPR loci structures reveals three elements (Fig 3.14):

1. Cas gene (Cas1 to 4)
2. Leader sequence
3. Repeat–spacer array

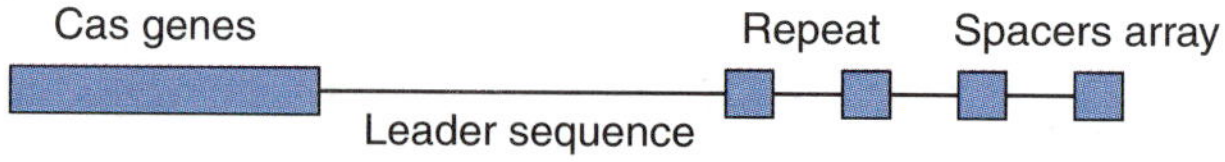

Figure 3.14 CRISPR-Cas system.

The arrangements of these three elements is not constant. It may be altered or even several CRISPR sequences may be present in a single genome.

In 2005, it was observed that some CRISPR spacers are derived from phage DNA or from plasmids, which thereby means that the spacers are DNA fragments from viruses that have previously attacked the cell. This also attests the possible role of CRISPR-Cas system in adaptive immunity.

Subsequent research led to the role of RNA transcripts of spacers on target recognition, the mechanism which is analogous to RNA interference system found in eukaryotes. This RNA interference hypothesis leads to mechanism of action of different CRISPR-Cas subtypes and predictive function and their proteins.

In 2008, researchers identified a complex of Cas proteins (called Cascade), which in *E. coli* cuts the CRISPR RNA precursor into mature spacer containing RNA molecules called CRISPR RNA (crRNA). This remained bound to the protein complex.

Cas9 is a protein, endonucleases coded by Cas gene. It serves as a molecular scissors. It has two small crRNA molecules and tracer RNA (trans- activating crRNA).

Researchers subsequently reengineered Cas endonuclease into two component system. This was achieved by fusing two RNA molecules into a "single guide RNA" which when combined with Cas9 could identify and cleave DNA target specified by guide RNA.

CRISPR-Cas9 System

In short, this system has

1. Cas9—endonuclease serves as molecular scissors
2. Spacer DNA—codes for crRNA
3. Tracer RNA—holds crRNA in place
4. Tracer RNA + crRNA = Guide RNA
5. Guide RNA + Cas9 protein = CRISPR-Cas system

With the help of CRISPR-Cas system, desired gene corrections/replacement can be done with accuracy. The genome editing in human cell was successfully done with CRISPR-Cas system by Feng Zhang and George Church.

In July 2019, CRISPR gene editing was used on experimental basis to treat a patient of sickle cell disease in Tennessee, USA.

Future

- CRISPR technology can be successfully applied in food and farming industry to obtain probiotic cultures. It can even help in immunising industrial cultures of products like "yogurt" against infections.
- It can be used to change mosquitos so that they will not transmit the malarial parasite.
- It can be used
  - To make biofuel
  - To modify plant species for better yield.
  - For revival of extinct species.
  - To understand gene–disease relationship.
  - For discovery of potentially important anomalies.

CRISPR technology has also emerged as a diagnostic tool. With only 250 pg of RNA as an input, diagnostic test can be done.

Newer diagnostic applications of this technology:

1. Electrical CRISPR chip
2. Microfluidic CRISPR biosensor

In true sense it has revolutionised the field of genetics.

HUMAN GENOME PROJECT

It is one of the most exciting and widely publicised project in the history of biomedical research. It was initiated in October 1990, with three major goals:

1. Genetic marker map
2. Physical map
3. Entire 3 billion base pairs sequence of the human genome.

The marker map was completed many years ago. It includes many thousand polymorphisms such as

a. RFLP^S—Restriction fragment length polymorphisms
b. STRP^S—Short tandem repeat polymorphisms
c. VNTR^S—Variable number of tandem repeats

Among these, useful polymorphisms are placed at an interval of 1 cM (centimorgan)—a distance between 2 loci, named after TH Morgan). In addition about 4 million SNPs (single nucleotide polymorphisms) have been found in the entire genome. They have lower mutation rate than polymorphisms. They are amenable to computerised automated processing and hence useful in human genetic mapping. The second goal of physical map of known STS (sequence tagged sites) placed at 100 KB intervals through the entire genome was also complete. These were especially useful in positional cloning experiments.

The third and the most important goal was to complete genome sequence. This was followed in both public and private sectors. In February 2001, both the groups announced completion of 90% sequencing with high degree of accuracy. This "rough draft" of genome contained most of our genes. Finally in 2003, complete and highly accurate sequence was announced i.e. completion of human genome project. This happens to be exactly 50 years after Watson and Crick described the structure of DNA.

With completion of Human Genome Project, the positional cloning is now possible with existing physical map. Numerous disease genes have been identified. Cloning of disease genes has improved genetic diagnosis and has also a potential to manifests gene products by recombinant DNA technology. This will provide improved health care and more specific drugs and gene therapy.

The complete genome sequence will provide ultimate genetic blueprint of the human being. Using similar technique of sequencing we can study medically significant bacteria and viruses. Sequencing of experimental organisms e.g. yeast, fruit fly, mice, rats can be done. Homology between sequence of these organisms and the human genome will offer us better understanding human disease genes.

In short completion of human genome project is the beginning of new era in the biomedical research.

Summary

Structure of Nucleic Acid: It consists of long chains of molecules called nucleotides; each having nitrogenous base, a sugar moiety and phosphorus molecule. The bases are of two types: Purine and pyrimidine. Purine bases are adenine and guanine, and pyrimidine bases are thymine, cytosine and uracil.

There are ***two types of nucleic acids***—DNA and RNA. DNA has sugar called *deoxyribose* and is mainly found in nucleus. RNA contains sugar called *ribose*, found mainly in nucleolus and cytoplasm.

Among pyrimidine bases, DNA has *thymine* and RNA has *uracil*; other bases are common.

DNA Structure (Watson–Crick Model): JD Watson, FHC Crick and MHF Wilkins suggested that *DNA consists of two chains of nucleotides. The chains are held together by hydrogen bonds. The chains have polarity and they are said to be anti-parallel. The bases of DNA pair with specificity. Adenine pairs with thymine and cytosine with guanine. During division, each strand builds up its complement, i.e. replication occurs.*

Types of DNA

- ***Unique sequences:*** These are genes coding for specific proteins.
- ***Satellite DNA:*** It consists of short tandem repeats coding for ribosomal and transfer RNAs. These are clustered in heterochromatic regions of chromosomes 1, 9, 16 and long arm of Y.
- ***Interspersed repetitive DNA sequences:*** These are of two types—(*i*) short interspersed repetitive sequences of about 300 bp called "*Alu repeats*"; and (*ii*) long interspersed repetitive sequences or *LI family* about 6000 bp long. Their function is not known. Both Alu and LI family sequences have been implicated in mutations in inherited human diseases.

Mitochondrial DNA: It is circular, designated as mtDNA. It codes for two types of ribosomal RNAs, protein subunits of some enzymes, e.g. cytochrome B, cytochrome oxidase.

Chromosome: "Solenoid model" proposed by Finch and Klug. It is thought that there may be several orders of DNA coiling such as—

- Primary coiling of DNA double helix.
- Secondary coiling around histone beads, i.e. nucleosomes.
- Nucleosomes undergo tertiary coiling to form chromatin fibres.
- Chromatin fibres form loops.
- Loops further coil to form chromosomes.

Triplet Code: The sequence of three bases that codes for an amino acid is called codon or triplet code. Four bases A, T, G and C make 43 = 64 combinations. In mutation, a particular code may be read as "*nonsense*"—no amino acid shall be coded

Continued

Summary—cont'd

or as "*missense*"—a different amino acid shall be coded and abnormal protein shall be produced.

UAA and UAG are called "*Ochre*" and "*Amber*"; they do not code for any amino acid.

Ribonucleic Acid (RNA): Types—messenger (mRNA), transfer (tRNA) and ribosomal (rRNA).

Messenger RNA and Heterogeneous Nuclear RNA (hnRNA): Primary product of transcription is hnRNA. It has "exon"-coding sequences and "introns"-non-coding sequences. The latter are removed and the former, i.e. exons, are spliced together to form mRNA. It then acquires methylated cap at 5′ end and poly A tail at 3′ end, and subsequently comes out of nucleus and reaches ribosomes in cytoplasm.

Transfer RNA (tRNA): It is single stranded and is folded forming hairpins. It has three loops, "Clover leaf Model"—*TC loop* (containing pseudouracil residue); *DHU loop* (containing dihydrouridine); and *Anticodon loop* (containing specific triplet complementary to codon of mRNA).

Ribosomal RNA (rRNA): Ribosomes occur as tiny particles in two sizes 60s and 80s. Ribosome has rRNA and ribosomal protein. They form string-like structure of polyribosome. Ribosome has two subunits. The rRNA in subunits of ribosome is highly folded. Its 3′ end has sequence complementary to mRNA ribosome binding site. The 5S rRNA in larger subunit of ribosome has sequence complementary to the TC loop sequence of tRNA. This allows binding of tRNA to ribosomes.

Transcription: A process of information transfer from DNA to mRNA; it involves the following:

- The two strands of DNA separate.
- Against each single strand of DNA, mRNA is synthesised. This occurs with complementary base pairing, i.e. cytosine pairs with guanine, thymine with adenine; however, adenine pairs with uracil.
- mRNA migrates to cytoplasm.

Translation: A process of translating information from mRNA into protein synthesis and involves the following—

- mRNA gets associated with ribosomes (sites of protein synthesis).
- mRNA serves as a template.
- Amino acid to be incorporated is activated by ATP. tRNA receives this activated amino acid. tRNA (with activated amino acid) has at its other end triplet complementary to mRNA. Thus, mRNA triplet through tRNA triplet picks up required amino acid.
- The ribosome with tRNA + amino acid moves along mRNA, linking amino acids in polypeptide chain.

Gene

- It is thought that there are 50,000–100,000 DNA sequences that code for RNA or protein products.
- The *structure of human globin gene* presents three exons (coding sequences) and two introns (non-coding sequences). At 5′ end, it has *CAT box*—70 bp upstream from the first exon and *TATA box*—located between CAT box and site initiation of transcription. At 3′ (downstream), it shows sequence AATAAA, i.e. signal for addition of poly-A tail.

Summary—cont'd

Generalisations:

- Genes are coding sequences split by intervening sequences.
- The exon–intron pattern of split genes is conserved during evolution.
- Alterations in exons occur slowly at the rate of 10^{-9} substitutions per codon per generation.

HTF islands: These are methylation-free HpaII tiny fragments called HTF (HpaII tiny fragments) islands. They contain one or more CpG dinucleotides near transcription initiation site at 5′ end of many genes.

Jumping genes: Discovered by Barbara McClintock, these are also called transposons; they can jump to and fro within a chromosome or to the adjacent one. In bacteria, they cause rapid spread of antibiotic resistance genes.

Pseudogenes: These are strikingly similar to functional genes, but are not transcribed.

Operon: Postulated by Jacob and Monod; the concept is that there are control genes such as regulator genes and operator genes. They regulate the action of structural genes, i.e. amount of protein produced by gene.

Mutation: It is change in sequence of genomic DNA. It can be:

i) Single base substitution or point mutation.
ii) Deletion or insertion: They may cause "frame shift".
iii) Chain termination mutations: May increase or reduce the amount of protein product.
iv) Splice mutations: Mutations involving excision of introns and splicing of exons.
v) Mutations of regulatory sequences.

Gene Mapping

Chromosome Mapping: Assigning a particular gene to a specific chromosome.

DNA Mapping: Relationship of flanking DNA sequence polymorphisms.

Pulsed Field Gel Electrophoresis (PFGE): Helps in separation of larger DNA fragments (2 million bp) and their mapping.

YAC Contigs: Yeast artificial chromosome (YAC) helps in cloning of larger DNA fragments, e.g. genes of cystic fibrosis, neurofibromatosis, dystrophin, etc.

Gene Cloning: Desired gene is inserted into the host cell through the vector (carrier molecule). The host cell starts to copy this gene and begins to produce protein molecules.

Gene Bank: It is a collection of artificial, recombined (recombinant) DNA molecules.

Indian National Gene Bank: Set up by National Bureau of Plant Genetic Resources (NBPGR). It has four complements:

i) Seed repository
ii) Tissue culture repository
iii) Cryopreservation facility
iv) Clonal repository

Recombinant DNA

i) It is artificially synthesised DNA.
ii) It is constructed with the help of restriction enzymes and vectors.

Vectors used are plasmids, phages, cosmids, YACs, etc.

Steps:

a. DNA is cut with restriction endonuclease enzyme.
b. Fragments are incorporated into vector.

Continued

Summary—cont'd

c. Recombinant vector is inserted into host cell.
d. These host cells are cultured to produce clones.
e. Selection of clones with proper DNA fragments.

Applications of recombinant DNA:

a. To understand molecular basis of diseases.
b. Production of insulin and other such products.
c. Production of proteins for diagnostic test, e.g. AIDS test.
d. Production of proteins for vaccination.

Polymerase Chain Reaction (PCR): It involves—

i) *Denaturation of DNA:* Converting it into single stranded (92–95°C)
ii) *Annealing:* Primer anneals at specific sites (55–65°C).
iii) *Primer extension:* Nucleotides are added, one at a time (72°C).

Analysis of PCR Product: Estimation of size, hybridisation, restriction enzyme mapping, cloning and sequencing.

i) ***Merits:*** (*a*) Sensitive; (*b*) small amount of DNA is required; (*c*) sample can be blood, bone chip, hair follicle, sperm or any other tissue.
ii) ***Demerits:*** (*a*) False positive reaction or (*b*) false negative reaction.

Applications of PCR: It is used in diagnostics, therapeutics, criminology, etc.

Southern Blot: In this, DNA is cleaved, the fragments are separated by electrophoresis and DNA is denatured (turned single stranded). This DNA is taken to nitrocellulose filter by blotting. Radioactive probe (P^{32}) is used to localise particular fragment. Probe hybridises with DNA fragments to be followed by autoradiography.

Northern Blot: In this, mRNA is isolated and run on an electrophoretic gel and is transferred to a filter. Its hybridisation with a radiolabelled probe allows determination of the size and quantity of mRNA transcript.

DNA Fingerprinting: Pioneered by Alec Jeffreys; it forms a powerful tool in criminology, comparing the DNA of the suspect with the sample collected at the site of crime. It is helpful in the diagnosis of genetic disorders on subclinical level. It helps in tissue matching before transplantation.

Gene Chip: These are devices as small as postage stamps, based on a glass substrate; these have many tiny cells that hold DNA from different human genes. This enables to carry out large number of genetic tests on a sample at one time. They have wide applications in pharmaceutical industry. They are more formally called microarrays.

Human Genome Project: It started in 1990 and was completed in 2003. There were three major goals of the project:

i) Genetic marker map
ii) Physical map
iii) Sequencing of all 3 billion base pairs

Completion of Human Genome Project marked beginning of new era in the field of biomedical research.

QUESTION YOURSELF*

1. What is operon?
2. What is mutation?
3. What is point mutation?
4. Is the following statement true—"A normal female is mosaic"?
5. What is nucleotide?
6. The DNA has all of the following bases except:
   a. Adenine
   b. Uracil
   c. Guanine
   d. Thymine
7. Which one of the following is not a pyrimidine base?
   a. Adenine
   b. Uracil
   c. Cytosine
   d. Thymine
8. What is triplet code?
9. Name different types of RNA.
10. What is hnRNA?
11. What is transcription?
12. What is translation?
13. What is gene?
14. What are jumping genes?
15. What are pseudogenes?
16. What is "frame shift mutation"?
17. What does PFGE stand for?
18. What is recombinant DNA?
19. What is plasmid?
20. All of the following are vectors used to carry DNA fragments except:
    a. Plasmid
    b. Yeast artificial chromosomes
    c. Cosmids
    d. cDNA
21. What does PCR stand for?

---

*See pages 258–259 for Answers.

4 Chromosomal Aberrations

COMPETENCY AN 75.1 **Describe the structural and numerical chromosomal aberrations.**
COMPETENCY AN 75.2 **Explain the term chimeras with examples.**
COMPETENCY AN 75.3 **Describe the genetic basis and clinical features of Prader–Willi syndrome, Edward syndrome, and Patau syndrome.**

LEARNING OBJECTIVES

At the end of this chapter the students should be able to understand
Genesis and effect of
- Numerical and structural aberrations
- Sex chromosomal abnormalities
- Down syndrome, Turner syndrome, Klinefelter syndrome

KEY WORDS

Aneuploidy, Trisomies, Hermaphroditism, Chimaera

The normal chromosomal complement in a male is 46,XY and in a female is 46,XX. Any deviation either in number or structure of the chromosomes is referred to as chromosomal aberration. In this context, it will be worthwhile to note certain terms.

Diploid: Refers to normal chromosome number in human beings, i.e. $2n = 46$.
Haploid: Refers to $n = 23$; it is found in gametes.
Polyploid: Multiple of n, i.e. 23 such as triploid = 69 or tetraploid = 92 chromosomes. These are referred as polyploidy.
Aneuploid: Any number that is not exactly a multiple of "n", i.e. 23, such as $2n - 1$ or $2n + 1$; the former is found in Turner syndrome (45,XO) and the latter in Down syndrome (47 chromosomes with 21 trisomy). It is to be noted that the monosomy involving autosomes is lethal. The only exception is a rare instance where an infant with monosomy 21 survived.
Genesis of "Aneuploidy": It results from non-disjunction during meiosis. This causes unequal distribution of chromosomes in daughter cells. Instead of a

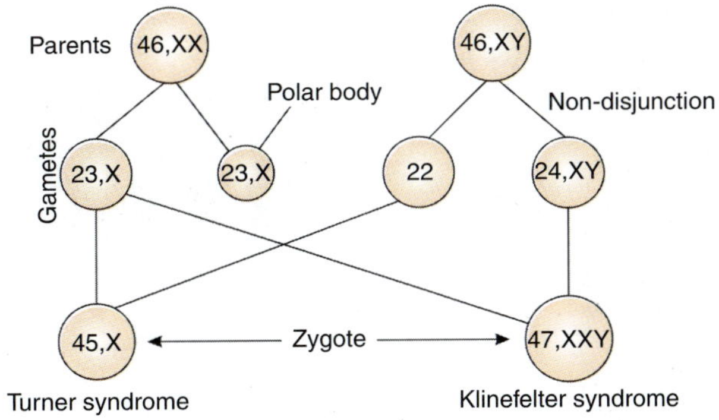

Figure 4.1 An outcome of non-disjunction during gametogenesis.

member of homologous chromosome pair, the pair goes to one daughter cell, and the other daughter cell is devoid of this chromosome. When this gamete with an abnormal number of chromosomes $2n - 1$ (22) or $2n + 1$ (24) combines with another normal gamete, the resultant abnormality is aneuploidy, like 45,XO (Turner syndrome) or 47,XXY (Klinefelter syndrome; Fig. 4.1). In the same manner, trisomies of autosomes are also formed, e.g. trisomy 21 or Down syndrome. Non-disjunction may occur at first or second meiotic division. It can also occur during cleavage (i.e. after zygote formation) resulting in trisomic and monosomic cell lines. An autosomal monosomy, however, does not persist but trisomy may continue.

STRUCTURAL ABERRATIONS **AN 75.1**

Structural rearrangements in chromosomes essentially result from breaks followed by reconstitution. The factors responsible for these are mainly (*i*) ionising radiations, (*ii*) chemical agents and (*iii*) viruses.

Structural aberrations are classified as under:

1. *Stable*, e.g. deletions, inversions, translocations, isochromosomes, etc.
2. *Unstable*, e.g. dicentric, ring chromosomes.

Among these, the aberrations that may be transmitted from parent to child include inversions or translocations.

Deletion

This involves loss of a part of chromosome. It is of two types (Fig. 4.2):

1. Terminal deletion
2. Interstitial deletion

Terminal Deletion

It involves a single break, and the terminal part of the chromosome is lost, e.g. Cri-du-chat syndrome.

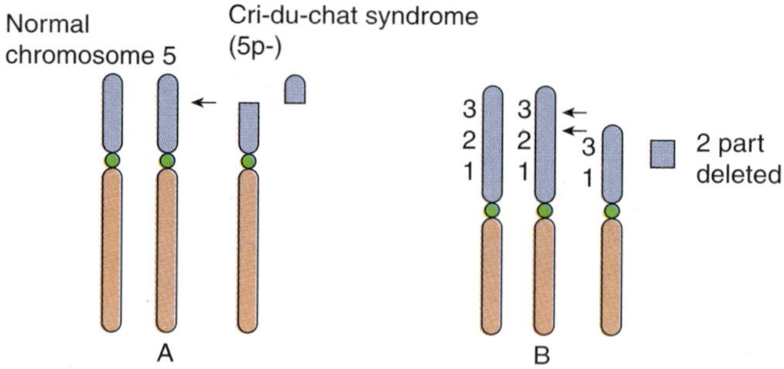

Figure 4.2 Chromosomal aberration—deletion: (A) Terminal deletion of P arm of chromosome. (B) Interstitial deletion of P arm of chromosome, e.g. Wilms tumour with aniridia (11p-).

Cri-du-chat syndrome or 5p-: This results from the deletion of the short arm of chromosome 5. It was first described by Lejeune and his associates. It is called Cri-du-chat syndrome because the cry of affected baby mimics mewing of a cat. Typical facial appearance, microcephaly, hypertelorism and anti-mongoloid slant of palpebral fissures form its classical features. Low-set ears, micrognathia are also found (Fig. 4.3).

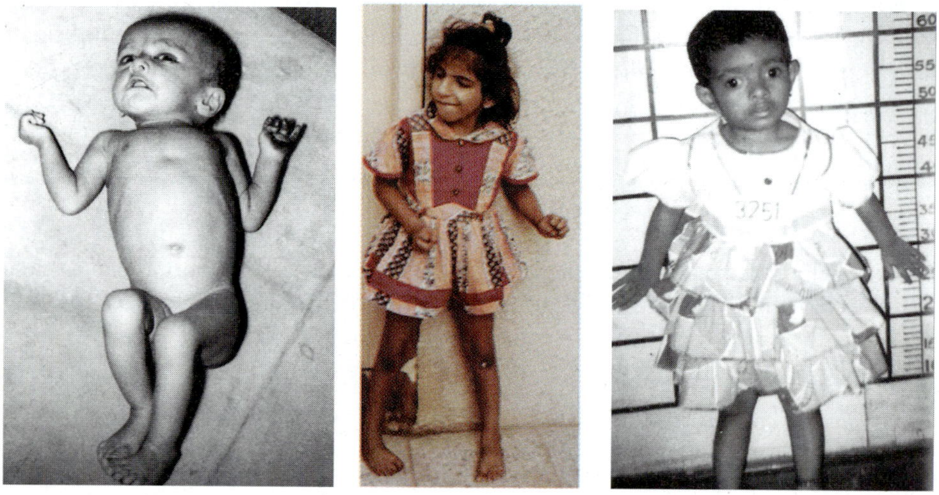

Figure 4.3 Photographs showing Cri-du-chat syndrome.

Interstitial Deletion

AN 75.3

It involves two breaks, and the intervening portion of the chromosome is lost, e.g. Prader–Willi syndrome (PWS), Wilms tumour with aniridia. They are called microdeletion syndromes.

Microdeletion syndromes

In the so-called microdeletion syndromes like in PWS, there is deletion of 3–4 million base pairs (3–4 Mb) of chromosome when this deletion is inherited from father. The phenotype presents short stature, hypotonia, obesity, small hands and feet with mild to moderate mental retardation and hypogonadism.

If the deletion is inherited from the mother, the child develops "Angelman syndrome", which is characterised by severe mental retardation, seizures and an ataxic gait. Now the question is—why there is difference. The portion of the chromosome 15 involved in both the syndromes is referred to as the "critical region". To explain this difference between paternal and maternal inheritance of the deletion (involving chromosome 15) leading to two different entities, we need to understand what is "genomic imprinting".

Genomic imprinting refers to differential activation of genes depending upon the parent from whom they are inherited. The transcriptionally inactive genes are said to be "imprinted". In the critical region of chromosome 15, several genes are transcriptionally active only on chromosome inherited from father, and they are inactive on the chromosome inherited from mother. Similarly, other genes in this region are transcriptionally active only on the chromosome inherited from mother and inactive on the paternal chromosome. This means, if the single "active" copy of these genes is lost due to deletion, then no gene product is produced, resulting into disease.

With the advent of high resolution banding (HRB), it is now possible to identify number of such deletions that were missed microscopically before HRB. Similarly, FISH techniques have made it possible to detect submicroscopic deletions known as microdeletions. There are often less than 5 Mb. For example, PWS was described in 1950; however, it was in 1981 that the precise location of the defect was identified with advanced banding techniques. In 50% cases, it involves deletion of paternal chromosome bands 15q, 11–q13. Microdeletion of the maternally-derived chromosome 15 produces genetically distinct Angelman syndrome. Table 4.1 shows microdeletion

Table 4.1 Microdeletion Syndromes

Syndrome	Chromosomal Deletion	Clinical Features
Angelman	15q11–13	Mental retardation, ataxia, seizures
Prader–Willi	15q11–13	Mental retardation, obesity, short stature, hypotonia, small hand and feet, typical facies
Miller–Dieker	17q13.3	Lissencephaly, characteristic facies
Wilms tumour with aniridia	11q13	Mental retardation, aniridia, predisposition to Wilms tumour, genital defects
Rubinstein–Taybi	16q13.3	Mental retardation, characteristic facies, vertebral abnormalities, pulmonary stenosis, "butterfly" vertebrae
Langer–Giedion	8q24	Characteristic facies, sparse hair, exostosis, mental retardation
Smith–Magenis	17q11.2	Mental retardation, hyperactivity dysmorphic features, self-destructive behaviour

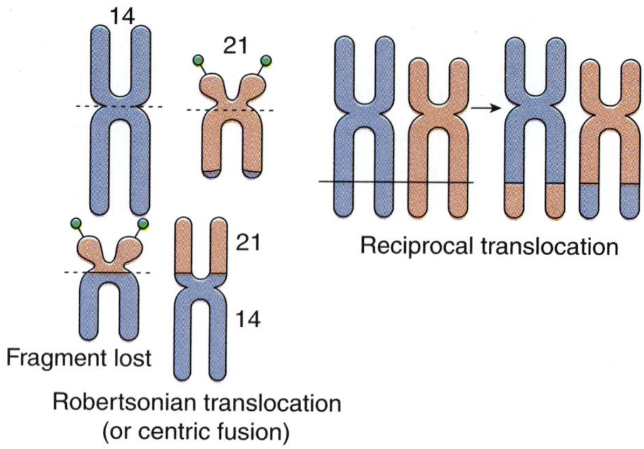

Figure 4.4 Types of translocations.

syndromes; however, some of these may be caused by single gene mutations in the chromosome regions.

Translocation

They are of two types (Fig. 4.4), which are described as follows:

1. **Robertsonian translocation:** This involves two acrocentric chromosomes, for example, D/G translocation. The short arm of a D group chromosome (13–15) fuses with the short arm of a G group chromosome 21. The fragment formed by their fusion is lost. This process is also called *centric fusion.* This type of translocation is found in about 4% of Down syndrome cases. Almost 50% of such translocation Down syndrome cases have parents as translocation carriers (balanced translocation). Another 50% account for a de novo event in the baby. Recurrence risk is high in the former and low in the latter.
2. **Reciprocal translocation:** In this, there is an exchange of chromosome material distal to breaks, and it involves non-homologous chromosomes. This amounts to a balanced translocation, and no chromosome material is lost. This, however, leads to the production of abnormal gametes presenting an unbalanced chromosomal complement, which in turn results in either spontaneous abortion or a baby with congenital malformations. In short, in case of repeated spontaneous abortions or a child with unbalanced translocation one can think of parents/couple having a balanced translocation carrier state.

Insertion

It is a rare non-reciprocal type of translocation that involves three breaks. A fragment is transferred from a chromosome to a non-homologous chromosome. Two breaks release the fragment from one chromosome and one break occurs in another chromosome to admit this fragment (Fig. 4.5).

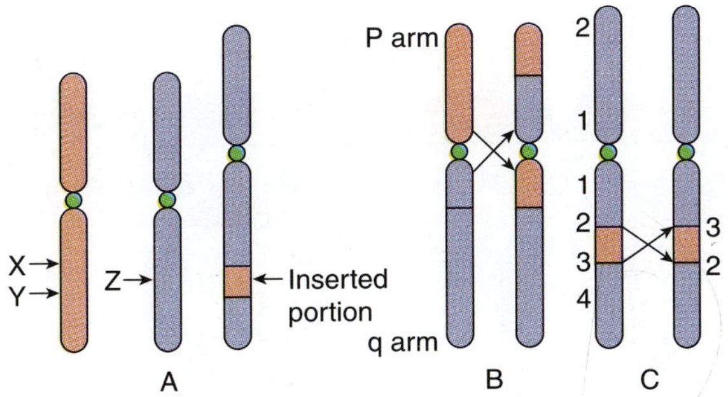

Figure 4.5 Structural aberrations in chromosomes: (A) Insertion, (B) Pericentric inversion, (C) Paracentric inversion. X, Y, Z in (A) indicate sites of break.

Inversion

It is of two types—*pericentric* inversion and *paracentric* inversion. Inversion involves two breaks along the chromosome. In pericentric inversion, both the arms p and q are involved, while in paracentric inversion only one arm either p or q is involved. Inversion does not give rise to abnormal phenotype in that individual. However, during meiosis abnormal gametes are formed giving rise to abnormal progeny.

Isochromosome

This involves abnormal split along the centromere leading to separation of arms. For example, i (Xq), i.e. isochromosome X (Fig. 4.6). It is found in some of the Turner syndrome patients.

Ring Chromosome

It involves two breaks at the terminal portions of the chromosome followed by fusion of the cut ends. This is found in about one-fifth of the cases of Turner syndrome (Fig. 4.7).

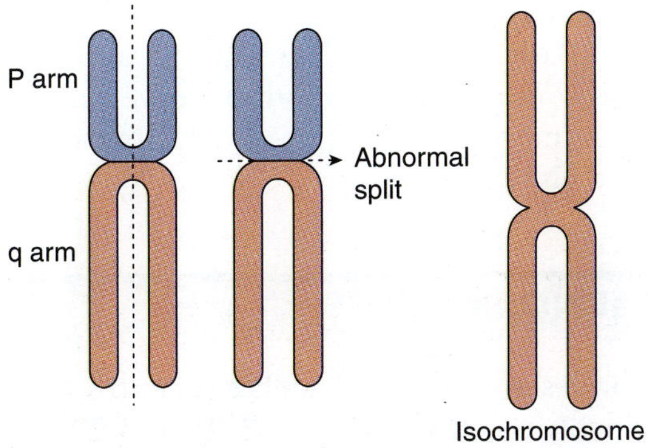

Figure 4.6 Formation of isochromosome.

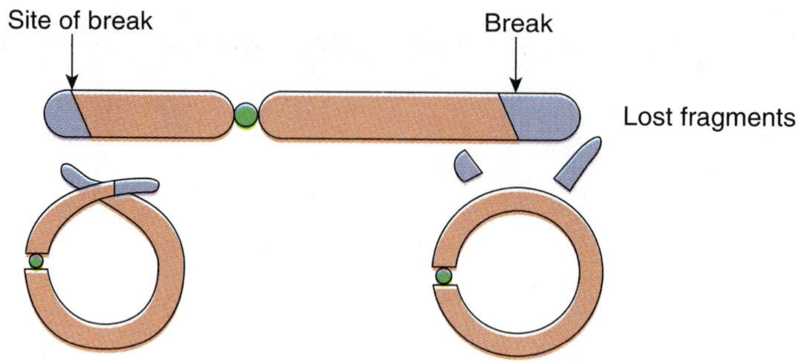

Figure 4.7 Formation of ring chromosome.

Factors Playing Role in Chromosomal Aberrations

1. **Maternal age:** Advanced maternal age (above 35 years) is one of the significant factors associated with Down syndrome. It is believed to be responsible for non-disjunction during meiosis I. This results in trisomy 21 (Down syndrome). Some studies indicate a possible role of late paternal age in the aetiology of Down syndrome.
2. **Non-disjunction gene:** The possibility of such a gene in human beings is being thought over. "Non-disjunction gene", however, occurs in other organisms. It may be responsible for non-disjunction in humans too.
3. **Radiation:** In 1977, Uchida presented data showing a correlation between radiation and non-disjunction in experimental animals. Various studies have indicated that radiation certainly increases frequency of Down syndrome.
4. **Chromosomal abnormality:** A balanced translocation in parents may result in an offspring with chromosomal aberration.
5. **Autoimmune disorders:** Though their precise role in the pathogenesis of non-disjunction is not very clear, it is believed that there exists a correlation between them. An association of high titre of thyroid autoantibody in mothers and Down syndrome in their children indicates the role of autoimmune disease in non-disjunction.

Now, we shall consider a few classical clinical syndromes presenting chromosomal aberration. Basically, this may involve an autosome or sex chromosome. Among several conditions identified so far, we shall consider only those that are relatively common. Table 4.2 shows chromosomal abnormalities in some well-recognised clinical patterns.

AUTOSOMAL ABNORMALITIES AN 75.1

Autosomal monosomies are fatal and such conceptus ends in abortion. Autosomal trisomies involving chromosome 13, 18 and 21 form well-recognised clinical entities and have been described in following pages.

Table 4.2 Chromosomal Abnormalities in Some Well-Recognised Clinical Patterns

Syndrome	Chromosomal Abnormality	Clinical Manifestations
Trisomy		
Down syndrome	Trisomy 21	Mental retardation, hypotonia, simian crease and characteristic facies
Edward syndrome	Trisomy 18	Mental and motor retardation, micrognathia, "rocker bottom feet", congenital heart disease
Patau syndrome	Trisomy 13	Mental retardation, microcephaly, microphthalmia, cleft-lip/palate, polydactyly
	Trisomy 8	Clinodactyly, other skeletal deformities, strabismus, moderate mental retardation
Deletions		
Wolf–Hirschhorn syndrome	4p-	Mental retardation, epilepsy, cleft lip/palate, coloboma, hypospadias
Cri-du-chat syndrome	5p-	Mental retardation, microcephaly, hypertelorism, cry like mewing of cat
De Grouchy syndrome	18q-	"Carp-mouth", mental retardation, abnormal ears and tapering fingers
De Grouchy syndrome	18p-	Mental retardation, dental decay, ocular and CNS abnormalities
Ring chromosome Anti-mongolism	21r	Anti-mongoloid slant of eyes, hypertonia, micrognathia, growth retardation and skeletal abnormalities

Down Syndrome (Trisomy 21, Mongolism)

It was first identified by Langdon Down in 1866. However, the chromosomal defect was unidentified till 1959. In this year, Lejeune and his associates found that patients with Down syndrome have 47 chromosomes instead of the normal 46. The extra chromosome was identified from the "G" group. It was designated as chromosome 21, a small acrocentric chromosome.

Langdon Down
(*Attributed to:* St. George's University of London.)

Clinical features

Mental retardation forms one of the predominant features in Down syndrome. The IQ level ranges between 25 and 50. Other features include small stature, hypotonia of muscles and brachycephaly with flat occiput.

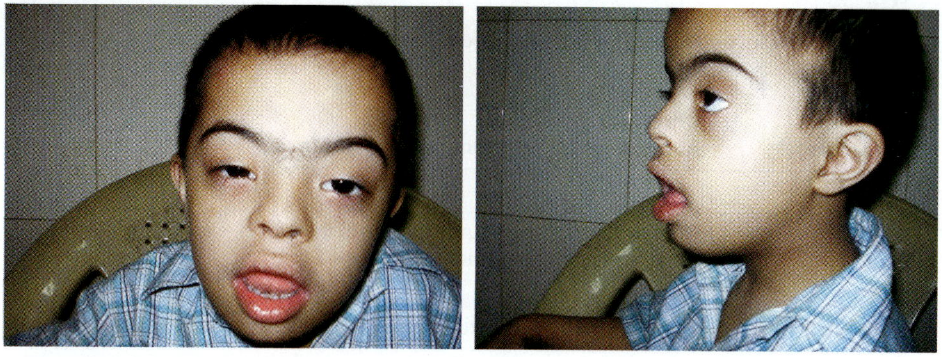

Figure 4.8 A boy with Down syndrome (trisomy 21). Note the epicanthal folds, depressed nasal bridge, low-set ears, open mouth.

The ears are low set and malformed, and the eyes show epicanthal folds producing a characteristic mongoloid slant; there may be nystagmus and the iris shows speckles. The flat nose presents a low nasal bridge (Fig. 4.8). The mouth is often open with tongue protruding. The tongue may be furrowed. The palate is often high arched, and the dentition may be delayed. Hands are short and broad, and there may be clinodactyly (incurving) of the little finger. Cardiovascular defects are also found in about one-third of the cases.

Dermatoglyphics

Simian crease forms one of the classical features. It is found in about 50% of Down syndrome cases. There may just be a single crease on the fifth finger. Axial triradius may be in the centre of palm in 85% of cases. There is often a wide gap between the first and second toe. About 50% patients show a hallucal dermal pattern as a tibial arch.

Cytogenetics

In almost 95% cases, there is trisomy 21 (Fig. 4.9). About 4% of the individuals show translocation, t (14q21q). Long arm of chromosome 21 is translocated to long arm of chromosome 14. In these patients having translocation, the number of chromosomes is 46, although they are trisomic for 21 chromosome. In about 1% cases, chromosomal complement is 46/47, i.e. they have mosaicism. They show two cell lines, a normal cell line of 46 and an abnormal cell line of 47 chromosomes (with trisomy 21). These patients (mosaics) are less severely affected. Mental retardation is relatively lesser as compared to a typical trisomy 21.

Risk of Down syndrome

Incidence of Down syndrome in the population is 1 in 800. In Israel, it is 1 in 400; in Malaysia, it is 1 in 500. This is probably related to girls' early age of marriage. In Israel, girls are married off at 8–9 years. Possibly the physical

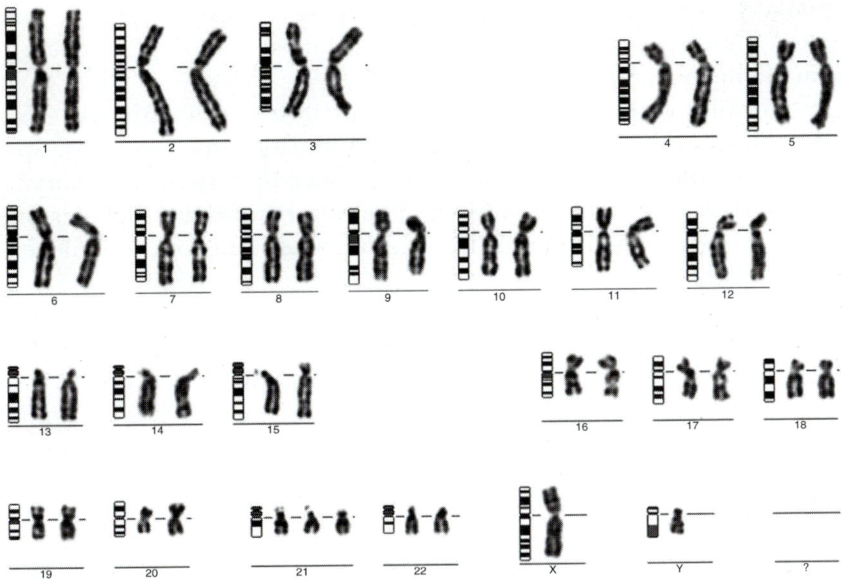

Figure 4.9 Karyotype of Down syndrome patient showing trisomy 21 (arrow).

and mental trauma they undergo may be contributing to high incidence (Survey by Mathru Mandir, Chennai, India, 1998). To calculate the risk to a mother of having a Down baby is a problem of genetic counselling. It depends upon a number of factors:

1. Maternal age.
2. Does the couple already have a baby with Down syndrome?
3. What is the karyotype of the baby (typical trisomy 21 or translocation)?
4. Is one of the parents a translocation carrier?

Prenatal diagnosis of the condition can be made with the help of chorionic villus biopsy or by amniocentesis.

Trisomy 18 or Edward Syndrome

AN 75.3

It was described by Edwards in 1960. It is also called E-trisomy. It is the second most common autosomal trisomy with prevalence of nearly 1 in 6000 live births. About 95% of the foetuses abort, only 5% of trisomy 18 conceptions survive to term. Trisomy 18 patients have prenatal growth deficiency. They have characteristic facial features and limb abnormalities clenching the diagnosis. Those who are born do not live beyond few months. Few may survive to about 15 years. E-trisomy presents with mental retardation and failure to thrive. Patients present with hypertonia, prominent occiput, receding jaw, low-set malformed ears. Ears may be small with unravelled helices; mouth is small. They have short sternum, clenched fists and rocker-bottom feet.

Congenital heart defects, such as ventricular septal defect (VSD), may be present. Other significant congenital anomalies are omphalocoele, diaphragmatic hernia and at times spina bifida.

High mortality rate in this condition is attributed to combined effect of factors such as aspiration pneumonia, predisposition to infections, apnoea and also congenital heart defects. Those trisomy 18 patients who survive infancy show developmental disabilities. The degree of delay of milestones is much more pronounced than in Down syndrome. Most of the children are unable to walk.

Cytogenetics

About 95% of babies with Edward syndrome present with complete trisomy 18. A small percentage shows mosaicism. Maternal age has a significant effect. Studies have indicated that nearly 90% cases among the patients of trisomy 18 have maternally-derived extra chromosome.

Trisomy 13 or D Trisomy or Patau Syndrome AN 75.3

It was first identified by Patau and his colleagues. The life span of patient is very much like that of trisomy 18. About 95% of the live born babies die during infancy. Those who survive infancy show significant growth retardation and severe mental retardation.

Clinical features

Clinically, it presents sloping forehead, hypertelorism, microphthalmia, coloboma iridis and postaxial polydactyly. Cleft lip, cleft palate are often present (Fig. 4.10). Facial cleft may also be seen in some patients. Congenital malformations involve cardiovascular system and urogenital system such as bicornuate uterus and polycystic kidneys. Severe central nervous system malformation such as holoprosencephaly may be seen occasionally. Cutis aplasia, a scalp defect on the posterior side along the occipital bone, may be present.

Cytogenetics

Nearly 80% of the trisomy 13 patients show an extra chromosome 13; however, others have trisomy involving only the long arm of chromosome 13 translocated. The risk of Patau syndrome increases with advanced maternal age, as in other trisomies (i.e. trisomy 18 and trisomy 21). About 95% of the trisomy 13 conceptions end up in spontaneous abortions.

SEX CHROMOSOME ABNORMALITIES AN 75.1

These may be presented in the form of trisomy XXY and XYY showing male phenotype, or monosomy involving X chromosome such as 45,X showing a female phenotype. Mosaicism involving X chromosome is more frequent

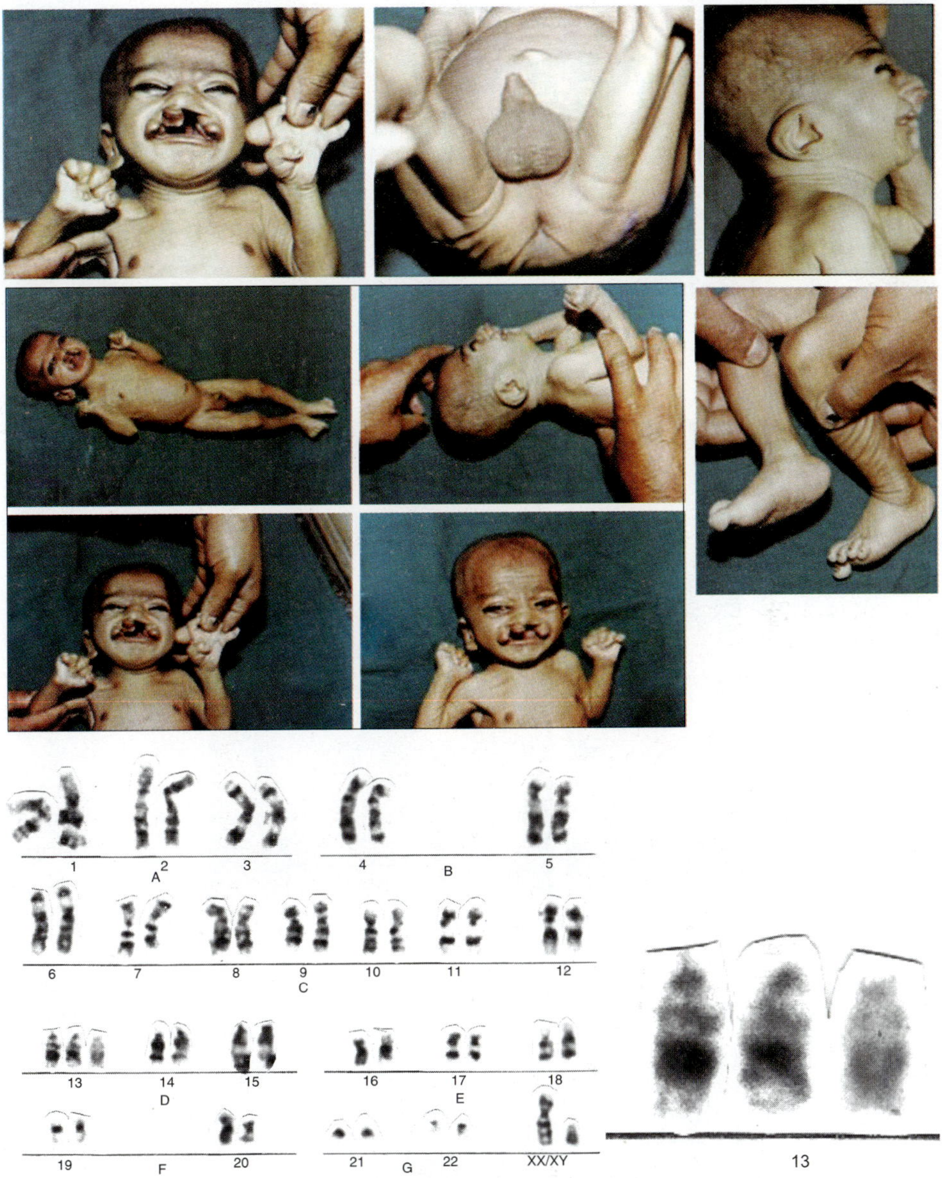

Figure 4.10 Patau syndrome (trisomy 13).

than seen in autosomes. About 50% of Turner syndrome patients and 15% of Klinefelter syndrome patients show mosaicism. Let us consider details of these two sex chromosome syndromes.

Turner Syndrome

It is also referred to as X monosomy. It was first described by Turner in 1938. However, the precise nature of cytogenetic abnormality was identified in

1959 by Ford et al. at Harwell. They demonstrated 45,X karyotype in Turner syndrome patients.

Clinical features

The phenotype in these patients is female. They have a short stature, webbing of neck, and cubitus valgus, i.e. reduction in the carrying angle at elbow (Fig. 4.11). Among other features, these patients have a low posterior hair line (Fig. 4.12) broad chest with widely spaced nipples. They have a high arched palate, lymphoedema over feet. There may be some of the following

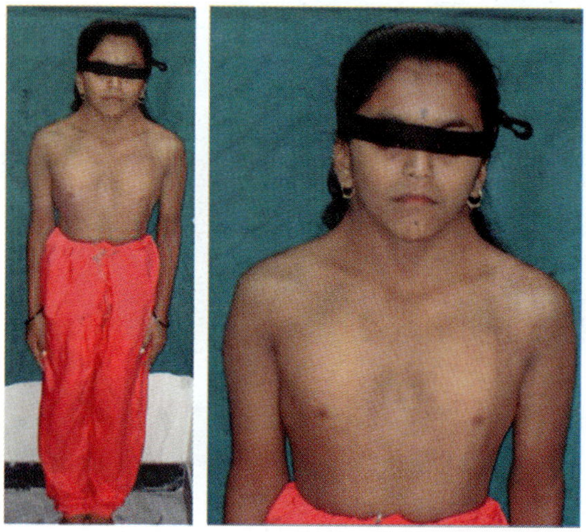

Figure 4.11 Turner syndrome case showing short stature and webbed neck.

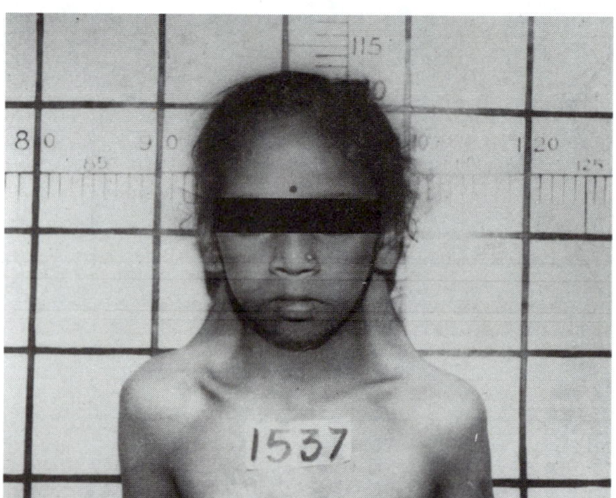

Figure 4.12 Note the webbing of neck in Turner syndrome patient.

congenital malformations involving various systems. In cardiovascular anomalies, there may be coarctation of aorta or VSD. In the urinary system, there may be horseshoe kidney, renal hypoplasia, or aplasia or duplication of ureters, etc. The genital system shows streak-like gonads (ovaries) consisting of connective tissue. There are no ovarian follicles. The uterus may be small. Secondary sexual characters do not develop. Primary amenorrhoea is usually present. Axillary and pubic hair are sparse. Normal breast development does not occur at puberty, and the external genitalia is of juvenile type.

Cytogenetics

The study reveals that about 60% of Turner syndrome patients show 45,X karyotype (Fig. 4.13). Others present a wide range of structural alteration involving X chromosome or mosaicism. The most common X alteration is in the form of 46, X, i (Xq). This is isochromosome involving long arm of X chromosome. Ring chromosome Xr is also not uncommon.

Investigations

1. Barr body examination reveals that patients are chromatin negative.
2. Dermatoglyphic study shows high total ridge count and distal axial triradius.
3. Karyotype shows 45,X in about 60% cases and others with structural abnormalities of X chromosome (Figs 4.14 and 4.15).

Their intelligence is normal or slightly less than normal. Failure to develop secondary sexual characters often brings them for consultation. Anabolic steroid therapy around 10–12 years of age helps them to gain height. Oestrogen administration helps development of secondary sexual characters.

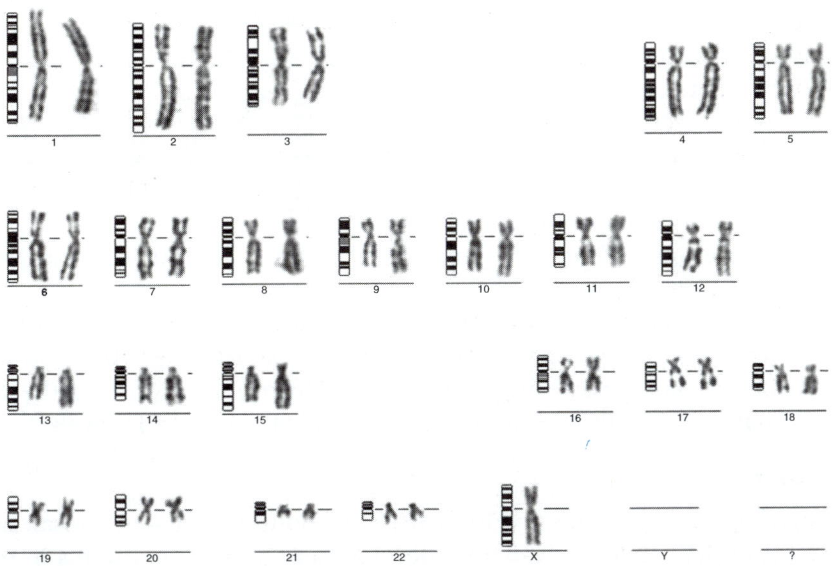

Figure 4.13 Karyotype of Turner syndrome showing only one X chromosome.

Polysomy X

It may be in the form of XXX, XXXX or XXXXX karyotype. Trisomy X presents with a female phenotype, which is almost normal. Usually, they are detected on examination and investigations for infertility and mental retardation. Somatic cells show two chromatin bodies. Among other polysomies (i.e. patients with four or five X chromosomes), patients develop severe mental retardation and have multiple physical defects (Fig. 4.16).

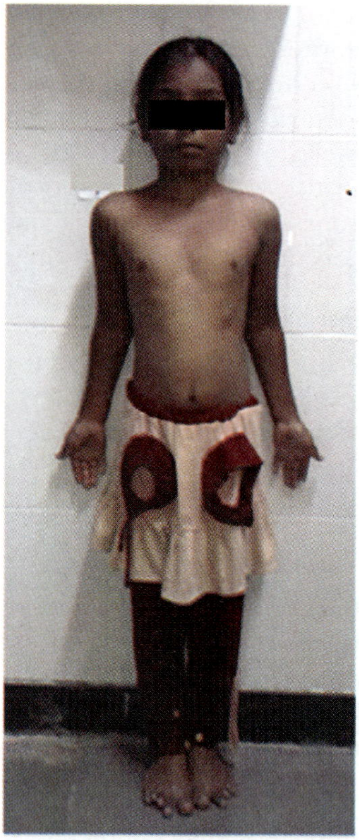

Figure 4.14 A female patient presented with short stature, wide gap between 1st and 2nd toe bilaterally, small 3rd and 4th toes, hyperconvex and upturned nails, revealed multiple cell lines on karyotyping, i.e. 46,XX(10%)/45,X(10%)/46,XX(iq)(60%)/46,XX (ter rea) (20%).

Klinefelter Syndrome

This condition was first described by Harry Klinefelter in 1942. The karyotype of these patients is 47, XXY. This was demonstrated by Jacobs and Strong in 1959. It presents a peculiar situation in which an individual with male phenotype is X-chromatin positive. This aroused interest in the investigators who subjected these patients to chromosome analysis.

Clinical features

Patients are tall, thin, eunuchoid. They have long legs and poorly developed secondary sexual characters. Testis are smaller in size; scrotum and penis may show hypoplasia. There is associated gynaecomastia in some cases. Pubic, chin, chest and axillary hair are absent or poorly developed. They have normal intelligence; however, verbal IQ is low (Fig. 4.17). Testicular biopsy shows hyalinisation of seminiferous tubules. Spermatogenesis is absent, and the patients are sterile. Barr body study, as seen earlier, shows chromatin-positive cells. Hormonal profile of these patients reveals low serum testosterone and higher FSH and LH levels.

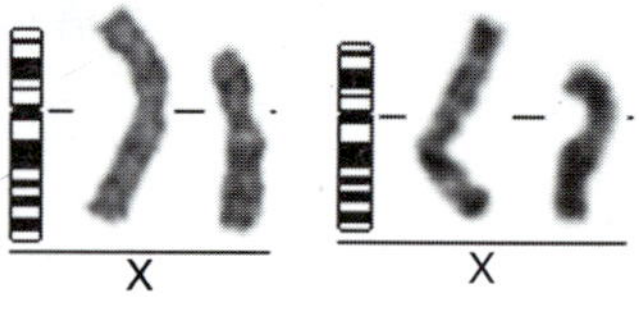

Figure 4.15 (A) Iso-Xq and (B) ter rea.

Cytogenetics

Karyotype is usually 47,XXY (Fig. 4.18). In about 15% cases, mosaicism is found, i.e. 46,XY/47,XXY. Testicular development and mental status in mosaics may be normal. In 60% patients, additional X chromosome is derived from

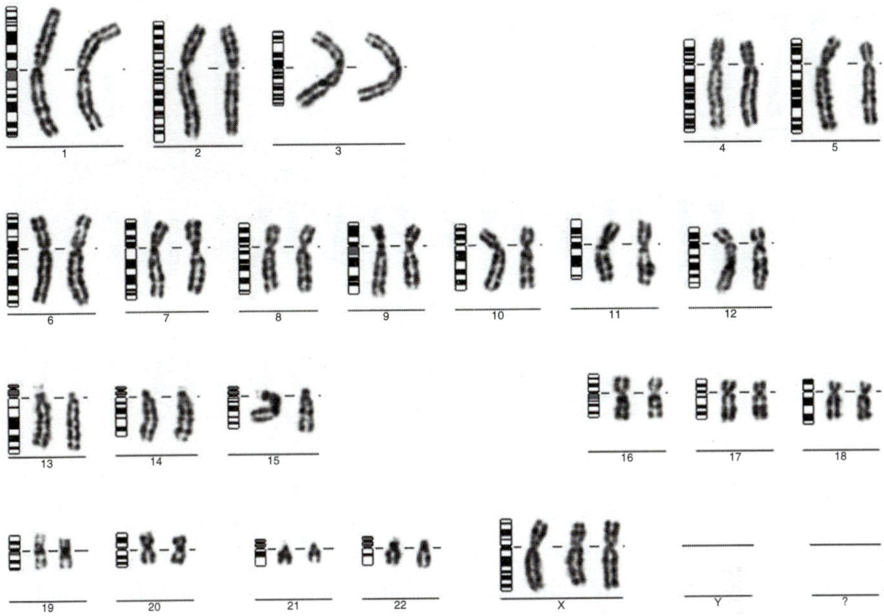

Figure 4.16 Karyotype: 47,XXX—Superfemale.

meiotic or postzygotic non-disjunction involving maternal X chromosome, i.e. $47,X^mX^mY$. In the remaining 40%, non-disjunction of X and Y chromosomes occurs during (first meiotic division of) spermatogenesis. This means, the chromosome complement is $47,X^mX^pY$. Variants of Klinefelter syndrome such as 48,XXXY or 48,XXYY or 49,XXXXY show additional X chromosomes with severe dysmorphism and mental retardation.

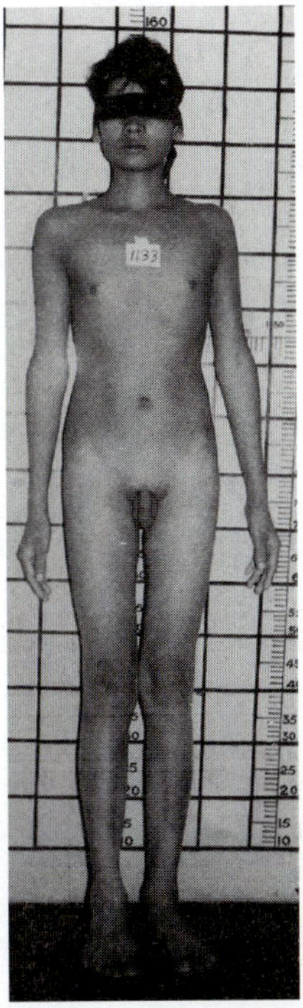

Figure 4.17 Photograph of a Klinefelter syndrome patient.

XYY Males

In this, an additional Y chromosome is found in a male phenotype. These individuals often show an emotional immaturity and impulsive character. This possibly associates them to anti-social behaviour. In fact in earlier studies, this karyotype was found with greater frequency among prisoners. It probably results from non-disjunction at second meiotic division producing YY sperm. Somatic cells of these individuals show two fluorescent spots on quinacrine dihydrochloride staining instead of a normal one.

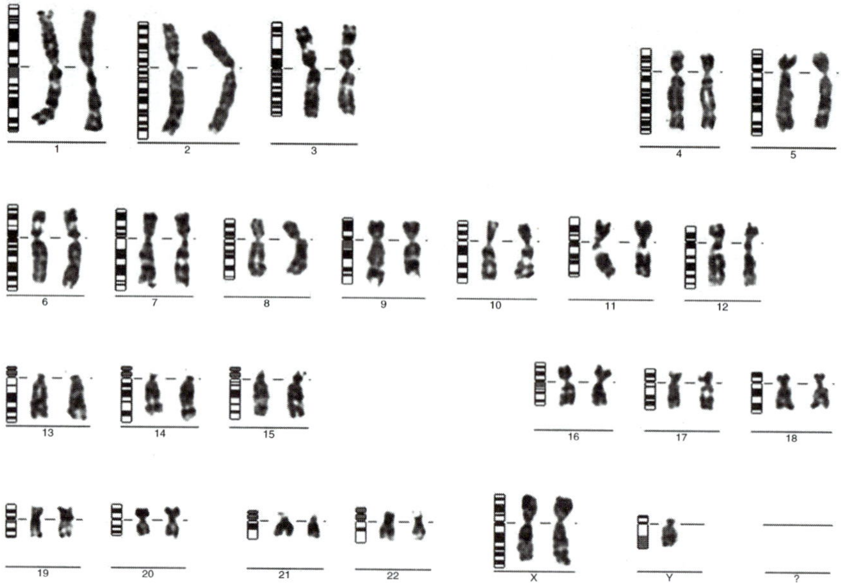

Figure 4.18 Karyotype of Klinefelter syndrome showing 47,XXY complement.

SEX DEVELOPMENT ERRORS WITH NORMAL CHROMOSOMES

In some individuals, it is difficult to assign sex because they have ambiguous genitalia. They do not have external genitalia clearly as a male or a female. In majority of them, there is only one type of gonad, either testis or ovary. Genital anomalies vary through a wide spectrum from hypospadias in male to hypertrophied clitoris in female. They may have normal sex chromosomes, but do present single gene defects or environmental factors leading to anomalies. Karyotyping forms an essential investigation in these patients for counselling.

True Hermaphroditism

It is rare but known. A true hermaphrodite has ambiguous genitalia of varying degree. It ranges from individuals who appear to be almost like a normal male to those who appear almost like a female. On exploration of gonad, one may find ovary on one side and testis on the other. There may be a mixture of testicular and ovarian tissue giving rise to ovotestis on both sides or on one side, while the other side shows a normal gonad. In these persons, one can expect mosaicism with two cell lines XX/XY. Some of them do show such mosaicism, but some however, show the XX complement.

Pseudohermaphroditism

As against a true hermaphrodite, a pseudohermaphrodite has only one type of gonadal tissue. A male pseudohermaphrodite possesses testis as gonads and shows XY chromosome complement. Female pseudohermaphrodites have an ovarian tissue and XX chromosome complement.

Female pseudohermaphroditism

It occurs with the frequency of about 1 in 25,000 births. The most common cause of female pseudohermaphroditism is congenital adrenal hyperplasia. It is inherited as an autosomal recessive disorder. It is characterised by a deficiency of cortical enzymes. As a result, the hormonal output from adrenal cortex is low. This, in turn, increases adrenocorticotropic hormone (ACTH) secretion from the pituitary. ACTH now causes adrenal hyperplasia. Hyperplastic adrenals elaborate androgens, which cause the masculinisation of female foetus leading to female pseudohermaphroditism. External genital examination shows hypertrophy involving clitoris; labia majora show rugosity and may even be partly fused.

Another event that may cause masculinisation of female foetus is the excess amount of sex hormones entering foetal circulation from mother. An overactive adrenal cortex of the mother or if the mother has received hormonal therapy, both events may lead to pseudohermaphroditism.

Male pseudohermaphroditism

It may be an outcome of any of the following errors:

1. Gonadal dysgenesis in embryonic development
2. Gonadotropins abnormality
3. Inborn errors in biosynthesis of testosterone
4. Androgen target cell abnormalities

Among these, the androgen insensitivity in target cells leads to what is commonly called testicular feminisation. Testicular feminisation syndrome is an X-linked disorder. In this, the patient has an XY chromosome complement. External genitalia shows female form, a blind vagina and there is no uterus or uterine tubes. Testicular tissue may be in abdomen or in inguinal canal. The receptor protein coded by allele at locus TFM forms a complex with testosterone. If this complex is not formed, then the hormone cannot enter the nucleus. Therefore, TFM has also been called a major sex determining gene in man. Testicular feminisation in its incomplete form may show clinically and genetically heterogenous types. They can be studied by analysis of androgen receptor-binding activity.

Role of Y Chromosome

The Y chromosome possesses H–Y antigen gene and male determining segment. The latter is responsible for development of testes. In turn, testes produce hormones responsible for masculinising effects. Experimentally, this has been proved by removal of testes from a foetal rabbit; the foetus developed into a female in spite of the XY chromosome constitution. Thus, Y chromosome necessarily accounts for maleness. It will not be inappropriate to mention about XX males at this stage. Males with XX karyotype (Fig. 4.19) occur

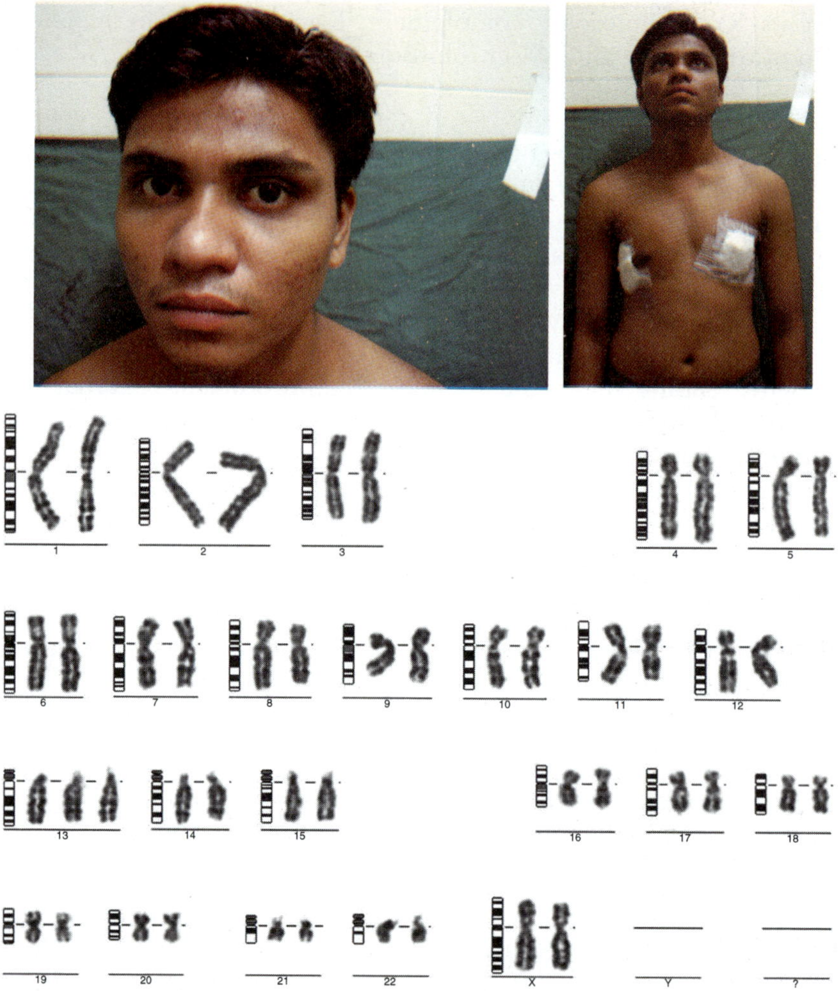

Figure 4.19 XX male, patient also had gynaecomastia that was operated.

with a frequency of about 1 in 15,000 male births. A possible explanation for XX male is as follows. They are probably XX/XXY mosaics, in whom the Y chromosome-bearing cell line has not been identified. This may hold true because XX males resemble Klinefelter syndrome. Another explanation is that during exchange between X and Y chromosomes in meiosis, male determining material associated with short arm of Y is translocated to X chromosome. Hence, despite the XX complement these individuals have a male phenotype.

CHIMAERAS

AN 75.2

So far we have seen what is a mosaic. Let us now consider another term "chimaera". Chimaera is an individual having two or more genetically different cell populations derived from more than one zygote. Originally, chimaera was named after a Greek mythological monster. It had the head of a lion, body of a goat and tail of a dragon. There are two types of naturally occurring chimaeras in man. Both are rare. These are (*i*) dispermic chimaeras and (*ii*) blood group chimaeras.

Dispermic Chimaera

AN 75.2

This is the result of double fertilisation. Two genetically different sperms (from different fathers) fertilise two ova. This results in the formation of two zygotes. If both contribute to the formation of an individual, it results in dispermic chimaera.

Blood Group Chimaera

AN 75.2

It can be formed by an exchange of cells across the placenta, between dizygotic twins. For example, the twins are non-identical, one of them has 80% XY cells and 20% XX cells. In the blood group analysis, many of his RBCs are of group B and few red cells belong to blood group A. The other twin member shows 80% XX cells and 20% XY cells. Her blood groups are—majority red cells show group A, while few red cells are of group B. Skin grafting usually takes up between identical twins, but in dizygotic twins it can take up satisfactorily if they are chimaeras.

Chimaeras have been produced in plants and in experimental animals too. To obtain chimaeras in animals is relatively difficult. However, they have been produced in mice. Eggs from pregnant mice are removed in the early stage of development. Two eggs from different strains are inserted in the presence of culture medium. After 1–2 days, the united eggs are transferred to the pregnant mouse to complete the development. Chimaeric mice can also be obtained by inserting mouse teratocarcinoma cells in mouse blastocyst (Fig. 4.20).

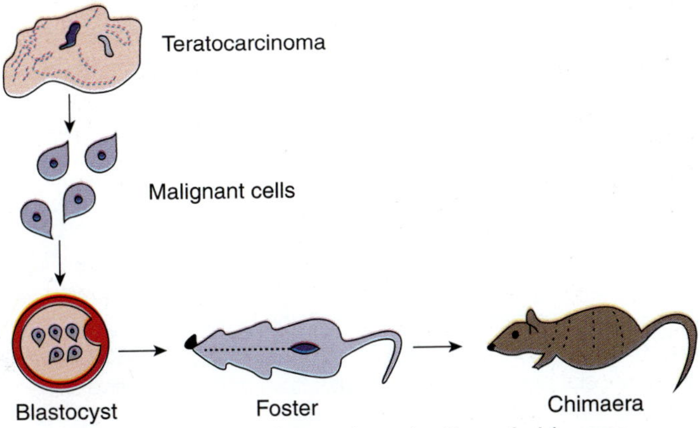

Figure 4.20 Experimental production of chimaera.

Summary

- **Normal chromosome number** in human beings is 46, it is called *diploid.*
- **Haploid (*n*),** i.e. 23 chromosomes; found in gametes.
- **Polyploid** refers to multiple of n, i.e. $3n = 69$ (triploid) or 92 (i.e. tetraploid).
- **Aneuploid** refers to any number that is not exact multiple of n or 23, e.g. $2n + 1 = 47$ chromosomes (Down syndrome) or $2n - 1 = 45$, a complement found in Turner syndrome 45,XO. *Cause* being "non-disjunction" at meiosis/gametogenesis.

Chromosomal aberrations can be numerical or structural aberrations: Monosomy—45,X (Turner syndrome); Trisomy—47,XX + 21 (Down syndrome).

- ***Trisomies:*** Trisomy 18, 13, 8 are known.
- ***Structural aberrations:*** 5p-, Cri-du-chat syndrome; 18q-, De Grouchy syndrome is known.
- **Structural aberrations could be:** (*i*) Deletions—terminal or interstitial deletion, e.g. Wilms tumour with aniridia; (*ii*) translocation—reciprocal or Robertsonian; (*iii*) insertion; (*iv*) inversion, either pericentric or paracentric; (*v*) isochromosome; (*vi*) ring chromosome.
- ***Factors responsible for chromosomal aberrations include:***
  1. Maternal age
  2. Non-disjunction gene
  3. Radiation
  4. Chromosome abnormality
  5. Autoimmune disorder/s

Autosomal Abnormalities

Down Syndrome—21 Trisomy (Langdon Down, 1866): MR with IQ between 25 and 50, brachycephaly, flat occiput, depressed nasal bridge, epicanthal folds, nystagmus, simian crease on hands, CVS defects, etc. *Karyotype:* 21 trisomy, translation, 14q 21q.

Trisomy 18 or Edward Syndrome: MR, prominent occiput, receding jaw, low-set ears, VSD, diaphragmatic hernia, spina bifida may be found. Ends in abortion or failure to thrive.

Trisomy 13 or Patau Syndrome: Sloping forehead, hypertelorism, microphthalmia, polydactyly, cleft lip, cleft palate anomalies of CVS, CNS and urogenital systems.

Sex Chromosome Abnormalities

Turner syndrome: Described by Turner in 1938. The patients have female phenotype, short stature, webbing of neck, cubitus valgus, high arched palate, amenorrhoea, poor secondary sexual characters. They may have coarctation of aorta, VSD, renal hypoplasia, etc. *Karyotype* is 45,XO or may present with isochromosome or ring chromosome involving X.

Polysomy X: Triplet X, i.e. 47, XXX; have MR and infertility.

Klinefelter syndrome: Harry Klinefelter described it in 1942. The patients are tall, thin, eunuchoid, having poor secondary sexual characters, hypoplastic gonad, azoospermia and low serum testosterone level. Gonadal biopsy shows hyalinisation of seminiferous tubules. *Karyotype* shows 47,XXY complement.

XYY males: They have male phenotype, have impulsive character and may be associated with antisocial behaviour. This results from non-disjunction at second meiotic division producing YY sperm.

Summary—cont'd

Hermaphroditism/Intersex

True hermaphrodite: It is rare. Gonads are testis on one side and ovary on the other side or may have ovo-testis. *Karyotype* shows mosaicism with XX/XY cell lines.

Female pseudohermaphroditism: Pseudohermaphrodite has only one type of gonad; female pseudohermaphrodites have ovaries and XX chromosomes complement. Common cause is congenital adrenal hyperplasia, with deficiency of cortical enzymes. There is masculinisation of female foetus, hypertrophy of clitoris and labial fusion. It is an autosomal recessive trait.

Male pseudohermaphrodite: Testicular feminisation syndrome is an X- linked disorder. External genitalia shows female form; however, vagina ends blindly, and there is no uterus or uterine tubes. The gonad is testis. They may be in abdomen.

Y chromosome: It possesses H-Y antigen gene and male determining segment responsible for testicular development. Foetal testes secrete testosterone that has masculinising effect on external genitalia.

Chimaera: Refers to an individual having two or more genetically different cell populations derived from more than one zygote. Naturally occurring two types are—

i) ***Dispermic chimaera:*** Two genetically different sperms (from two different men) fertilise two ova. Both zygotes contribute to form dispermic chimaera

ii) ***Blood group chimaera:*** Exchange of cells across placenta between dizygotic twins leads to blood group chimaera.

QUESTION YOURSELF*

1. Edward syndrome is:
   a. Trisomy 21
   b. Trisomy 18
   c. Trisomy 13
   d. Trisomy 8
2. Cri-du-chat syndrome is:
   a. Deletion involving short arm of chromosome 5
   b. Deletion involving long arm of chromosome 5
   c. Interstitial deletion of short arm of chromosome 11
   d. Deletion of terminal part of long arm of chromosome 11
3. What is aneuploidy?
4. Monosomy involving which chromosome is compatible with life?
5. What are the types of translocations?
6. What is Robertsonian translocation?
7. What is reciprocal translocation?
8. Why individuals with reciprocal translocation present with normal phenotype?
9. Why individuals with reciprocal translocation having normal phenotype produce abnormal offspring?
10. What is isochromosome?

11. Which one of the following syndrome patients exhibit webbing of neck?
    a. Klinefelter syndrome
    b. Down syndrome
    c. Turner syndrome
    d. Edward syndrome
12. Which one of following karyotype is found in Klinefelter syndrome patients?
    a. 45,XO
    b. 47,XXY
    c. 47,XXX
    d. 47,XYY
13. Which one of the following holds true about XYY males?
    a. They are highly intelligent
    b. They are impulsive and have criminal tendency
    c. They have short stature
    d. Their extra Y makes them more fertile
14. Match the following:

Condition	*Karyotype*
1. Turner syndrome	a. 47,XYY
2. Cri-du-chat syndrome	b. Trisomy 18
3. Criminal tendency	c. 45,XO
4. Edward syndrome	d. 5p-

15. What is true about male pseudohermaphroditism?
    a. They have 46,XX chromosome complement
    b. They have ovotestis as gonads
    c. They have testis as gonads
    d. They have 47,XXY chromosome complement
16. What is chimaera?

*See pages 259–261 for Answers.

Developmental Genetics 5

LEARNING OBJECTIVES

At the end of this chapter the students should be able to understand

- Factors influencing development
- Genes involved in the process of development
- Sex determination and differentiation

KEY WORDS

HOX genes, PAX gene, Zinc finger gene, Hydatidiform mole

Human development is a highly complex process with poor understanding in many areas. It is significantly influenced by both genetic and environmental factors. Under suitable environment, genes inherited from both the parents determine the fate of undifferentiated cell cluster derived from the zygote. By about 12 weeks, it acquires a recognisable human form. For the discoveries concerning genetic control of early embryonic development, Nobel Prize was awarded jointly to Edward B. Lewis, Christiane Nüsslein-Volhard and Eric F. Wieschaus in 1995.

In this chapter, we shall concentrate upon the genetic control of the physiological processes in the initial period of 12 weeks. In this respect, we deal with the branch called *developmental genetics.* The entire prenatal period is divisible into three periods—pre-embryonic, 30 hours to 19 days; embryonic, 4–12 weeks; and foetal, till birth. During pre-embryonic period, there is proliferation of the cells followed by formation of bilaminar and trilaminar disc. In embryonic period, craniocaudal and dorsoventral orientation of the embryonic disc is gained. Analogues of all major organ systems are laid down. During foetal period, further growth and development occurs in all the systems forming a viable human being.

FERTILISATION

It is a process in which male and female gametes fuse to form zygote. This occurs in Fallopian tube. About 200–300 million sperms are deposited in the female genital tract. Some of them ascend up reaching the site of fertilisation. Of these, some cross *corona radiata* and *zona pellucida*. One of these enters the cytoplasm of an oocyte. After this, the oocyte completes its second meiotic division, its nucleus now forms *female pronucleus*. By now sperm head transforms into a male pronucleus. The two pronuclei fuse to restore diploid number of chromosomes ($2n = 46$). The zygote then undergoes cleavage, i.e. a series of rapid mitotic divisions forming 16-cell stage, in about 3 days. It is called morula. Subsequently, it forms blastocyst. It consists of inner cell mass or *embryoblast* and outer *trophoblast*. The embryoblast then organises to form initially *bilaminar* and then *trilaminar embryonic disc*. The process is called gastrulation.

During the subsequent period, primitive streak is formed in the caudal half of the embryo. Endoderm, mesoderm and ectoderm (i.e. three germ layers) are formed. The period of 4–8 weeks is a critical period during which organogenesis occurs.

FACTORS INFLUENCING DEVELOPMENT

Knowledge regarding genetic factors influencing, maintaining and directing embryogenesis is limited. Studies in *Drosophila melanogaster* (fruit fly) led to identification of several genes that play a significant role in early developmental processes. These genes produce proteins called transcription factors. They control RNA transcription from DNA template by binding to specific regulatory DNA sequences. They form complexes that initiate transcription by RNA polymerase.

The transcription factors can switch genes "on and off" by activating or repressing gene expression. Some of these transcription factors, through specific genes, control fundamental embryological processes such as induction, segmentation, migration, differentiation, programmed cell death or apoptosis. This is mediated by growth factors and chemical agents known as *morphogens*. They stimulate the cell receptors and show *concentration gradient* across a structure such as a limb bud. In chick limb bud, fibroblast growth factor and bone morphogenetic protein have been identified as growth factors. The main morphogen involved in digit formation in a limb bud is identified as retinoic acid.

GENES INVOLVED IN DEVELOPMENT

There are three gene families so far identified in vertebrates that influence development. These are *Hox* genes, paired box genes and zinc finger genes.

Homeobox (Hox) Genes

The homeotic genes in *Drosophila* determine segment identity. Their mutations cause major structural abnormalities, for example, instead of antenna,

Table 5.1 Homeobox Gene Clusters Identified in Humans

hg	Chromosomal Location	Number of Genes
*Hox*1 (*Hoxa*)	7p	11
*Hox*2 (*Hoxb*)	17q	9
*Hox*3 (*Hoxc*)	12q	9
*Hox*4 (*Hoxd*)	2q	9

leg development occurs. The homeotic genes contain a conserved 180 base pair sequence called *homeobox*. This is the characteristic of the genes involved in spatial pattern control and development. The *Hox* genes are important transcription factors that specify cell fate and establish regional axis. Four homeobox gene clusters have been identified in humans (Table 5.1). Each cluster contains series of closely linked genes.

Mutations involving homeobox genes have not been identified in humans; possibly they are so devastating that they might be leading to abortion. Mutations of *Hox* genes in transgenic mice present with multiple and severe malformations involving skull and face.

Paired Box (Pax) Genes

The paired box is a highly conserved DNA sequence that codes about 130 amino acids. The genes that contain paired box are called *Pax* genes. They were first identified in *Drosophila*. They encode DNA-binding proteins. These proteins are transcription control factors and have an important role in development. So far, eight *Pax* genes have been identified in humans and mice. In mice, mutations involving *Pax*1, *Pax*3 and *Pax*6 lead to severe vertebral malformations, pigment abnormalities and small eyes, respectively.

In human beings, mutations involving DNA-binding domain of *Pax*3 cause Waardenburg syndrome, characterised by deafness, white forelock and heterochromia involving iris (Fig. 5.1). It is an autosomal dominant trait. Mutations in *Pax*6 lead to aniridia, i.e. absence of iris. Rearrangements involving *Pax*3 have been found in a childhood tumour, the alveolar rhabdomyosarcoma. These rearrangements are caused by translocation that interrupts *Pax* locus and forms a novel hybrid transcript.

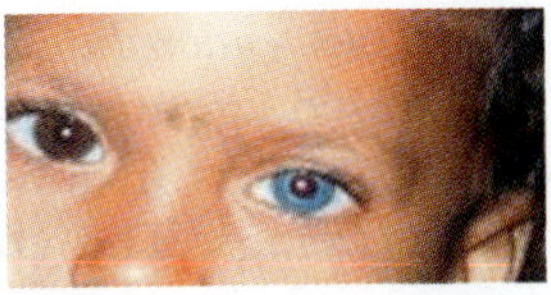

Figure 5.1 A patient of Waardenburg syndrome showing heterochromia.

Zinc Finger Genes

This describes a finger-like projection formed by amino acids, located between two cysteine residues, which forms a complex with zinc ion. Recently, it has been shown that an interruption involving a multiple zinc finger gene called *GL*13 on chromosome 7 causes Greig cephalopolysyndactyly. It is an autosomal dominant disorder. It is characterised by cranial and hand malformations, e.g. webbed digits. Mutation in Wilms tumour gene (*WT*1), located on 11p, presents with renal malignancy.

MULTIPLE MALFORMATION SYNDROMES

The mutations in transcription control genes provide an explanation for unrelated abnormalities in the multiple malformation syndromes. In embryogenesis, the concept of developmental field gives explanation accounting for these apparently unrelated embryonic primordia reacting to the genetic or environmental insult. The developmental field includes all organs or tissues having common embryological origin. Thus, an ectodermal dysplasia caused by single gene mutation may present with abnormalities involving teeth (enamel), hair, nails and sweat glands. Likewise in axial mesodermal dysplasia that occurs as a sporadic event in a family, organs of mesodermal origin such as kidneys, heart and vertebrae present an abnormal development. Similarly, VATER association features vertebral, anal, tracheo-oesophageal and rectal abnormalities.

Abnormalities in Waardenburg syndrome can be explained by abnormality involving neural crest cells. Thus, knowledge of embryology, genetics and dysmorphology helps us in getting rational explanation for congenital multiple malformation syndromes.

Hydatidiform Mole

In this abnormal conception, the placenta consists of a proliferating disorganised mass called hydatidiform mole. It may be partial or complete.

Complete hydatidiform mole

Here, the chromosome analysis reveals 46 chromosomes, but of exclusively paternal origin. Complete mole results from fertilisation of empty ovum (without nucleus) by either two sperms (dispermy) or by a single sperm that undergoes *endoreduplication*. An opposite situation may exist in which an unfertilised egg undergoes development. The process is called *parthenogenesis*. This occurs in lower animals (e.g. arthropods) and is not reported in human beings. The complete hydatidiform mole may undergo malignant change and give rise to choriocarcinoma. It is amenable to chemotherapy; if untreated may prove fatal.

Partial hydatidiform mole

The chromosome complement of the tissue from the partial mole reveals triploidy, i.e. 69 chromosomes. Using DNA polymorphisms, it is found that out of these, 46 chromosomes are paternal and 23 are maternal in origin. This can be due to dispermy (fertilisation by two sperms) or *endoreduplication*, i.e. duplication of haploid sperm chromosomes. Usually, it ends in abortion.

Role of Parental Chromosomes: It is observed in mice that when all the nuclear genes are derived from mother, the embryo develops normally but the extra embryonic (i.e. trophoblast) development is poor. In contrast, if all the nuclear genes are paternal in origin, then the embryo fails to develop; however, the trophoblast development proceeds unimpaired. The situation is comparable to what is observed in a complete and a partial hydatidiform mole.

SEX DETERMINATION AND DIFFERENTIATION

The sex of an individual is determined by Y chromosome. Presence of Y chromosome leads to male and its absence results in female development.

Though sex chromosomes are present from the time of fertilisation, differentiation into a male or a female does not occur till seventh week. The embryonic gonad and the genital duct system of Mullerian and Wolffian ducts are in indifferent stage. From seventh week onwards, the sequence of events initiated by testis determining factor (TDF) prompts development of indifferent gonad into testes.

It was shown in 1990 that testis determining factor/gene (TDF) is located on "p" (short) arm of the "Y" chromosome close to pseudoautosomal region. This gene is now said to be located in the sex determining region of the Y chromosome (SRY). Several observations indicate that the *SRY* gene is the primary factor determining maleness. It is supported by the following:

1. In XX males, *SRY* sequences are present. They are infertile phenotypic males with 46,XX karyotype.
2. The mutations involving *SRY* sequences are found in some XY females. They are infertile phenotypic females with 46,XY karyotype.
3. Tra.nsgenic XX mice that have a tiny portion of Y, containing *SRY* gene develop into males.

The close proximity of *SRY* gene to pseudoautosomal region means that occasionally it may be caught in recombinational event. This accounts for XX males. FISH studies show evidence of Y chromosome sequences at the distal end of the X chromosome.

The expression of *SRY* gene leads to development of medulla of indifferent gonad into the testis. Its Leydig cells produce testosterone. It promotes differentiation of Wolffian duct that forms elements of internal genitalia in males, i.e. ductus deferens, epididymis, seminal vesicle, ejaculatory ducts, etc. The masculinisation of the external genitalia is mediated by dihydrotestosterone. The latter is produced by the action of 5-a-reductase on the testosterone. The Sertoli cells of the developing testes produce a hormone called Mullerian inhibitory factor. This inhibits Mullerian ducts development and they regress.

In the absence of *SRY* gene expression, the cortex of an indifferent gonad develops into an ovary. The Mullerian ducts form internal genitalia. The external genitalia evolves into female form due to absence of dihydrotestosterone. An absence of testosterone also causes regression of Wolffian duct system.

INACTIVATION OF X CHROMOSOME

Out of the two "X" chromosomes in female, one undergoes inactivation to form Barr body. This is called lyonisation, named after Dr Mary F. Lyon who explained the X inactivation in her Lyon's hypothesis. For details refer to Chapter 2, Cytogenetics.

Summary

Developmental Genetics: It deals with genetic control of physiological processes in the initial period of development. The entire prenatal period is divisible into three periods—(*i*) Pre-embryonic period; (*ii*) Embryonic period; and (*iii*) Foetal period.

Factors Influencing Development: There are several genes that play a significant role in the early developmental process. They produce proteins called *transcription factors.* These transcription factors can switch genes "on and off" and control fundamental developmental processes such as induction, segmentation, migration, differentiation, apoptosis (programmed cell death). All these are mediated through growth factors and morphogens, e.g. main morphogen involved in digit formation in a limb is identified as retinoic acid.

Genes Involved in Development

Three gene families have been identified. *Hox, Pax* and Zinc finger genes.

(i) ***Homeobox (Hox) genes:*** Four homeobox gene clusters have been identified in humans—*Hox*1 (*Hoxa*) at 7p; *Hox*2 (*Hox*b) at 17q; *Hox*3 (*Hox*c) at 12q and *Hox*4 (*Hox*d) at 2q; these are their chromosomal locations. Mutations involving *Hox* genes are so devastating that they lead to abortion.

(ii) ***Paired box (Pax) genes:*** Pair box codes for about 130 amino acids. So far eight *Pax* genes have been identified. In humans, *Pax*3 mutations cause Waardenburg syndrome, an autosomal dominant (AD) trait, while *Pax*6 mutations lead to aniridia. *Pax*3 rearrangements have been observed in alveolar rhabdomyosarcoma.

(iii) ***Zinc finger genes:*** Recently, an interruption in multiple zinc finger gene *GL*13 on chromosome 7 has been observed in Greig cephalopolysyndactyly, an AD trait.

Multiple Malformation Syndrome

The concept of developmental field gives explanation accounting for apparently unrelated embryonic primordia reacting to the genetic and environmental insult, e.g. ectodermal dysplasia, a single gene mutation presenting with abnormalities involving teeth, hair, nails and sweat glands. Similarly, VATER association features vertebral, anal, trachea-oesophageal and rectal abnormalities.

Hydatidiform mole: An abnormal conception in which placenta proliferates into a disorganised mass called hydatidiform mole. It is of two types—

i) ***Complete hydatidiform mole:*** It shows 46 chromosomes on karyotyping, but all of them are paternal in origin. This results from fertilisation of an empty ovum (non-nucleated) by either two sperms (dispermy) or by single sperm that undergoes *endoreduplication*. In contrast, an unfertilised egg may undergo development, the process is called parthenogenesis.

ii) ***Partial hydatidiform mole:*** It shows triploidy, i.e. 69 chromosomes with 46 of paternal and 23 of maternal origin. It ends in abortion.

Summary—cont'd

Sex Determination and Differentiation

It is determined by Y chromosome at fertilisation; however, till seventh week the gonads and genital ducts are in *indifferent stage*. "p" arm of Y chromosome has testis determining factor (TDF). It is located in sex determining region of Y chromosome (SRY). SRY gene leads to development of medulla of indifferent gonad into testis. Leydig cells secrete testosterone that leads to differentiation of Wolffian ducts into internal genitalia of males, i.e. vas deferens, seminal vesicles, ejaculatory ducts, etc. The dihydrotestosterone causes masculinisation of external genitalia. In the absence of SRY gene, indifferent gonad forms ovary.

Inactivation of X chromosome: One of the X chromosomes in female becomes inactive to form Barr body. It is called *Lyonisation*.

QUESTION YOURSELF*

1. What are Hox genes?
2. What are Pax genes?

*See page 261 for Answers.

6 Modes of Inheritance

COMPETENCY AN 74.1 **Describe the various modes of inheritance with examples.**
COMPETENCY AN 74.2 **Draw pedigree charts for various types of inheritance and give examples of diseases of each mode of inheritance.**
COMPETENCY AN 74.3 **Describe the multifactorial inheritance with examples.**
COMPETENCY AN 74.4 **Describe the genetic basis and clinical features of achondroplasia, cystic fibrosis, haemophilia, and Duchenne muscular dystrophy (DMD).**

LEARNING OBJECTIVES

At the end of this chapter the students should be able to understand

- Mendelian inheritance (Single gene disorder)
- Basis of autosomal dominant and recessive traits
- Basis of sex/X-linked inheritance

KEY WORDS

Autosomal dominant, Autosomal recessive, X-linked recessive, Genetic isolates, Fragile X, Pleiotropy

We realise that the genetic disorders can be classified into three categories:

1. **Chromosomal disorders:** These occur because of abnormality in the number or structure of chromosomes, e.g. trisomy 21, monosomy of X or Cri-du-chat syndrome (5p-), etc.
2. **Single gene disorders:** These are due to a single mutant gene. They are also called Mendelian disorders. These have four basic patterns of inheritance—autosomal dominant, autosomal recessive and X-linked inheritance either recessive or dominant.
3. **Multifactorial inheritance:** In this, the disorder is a result of interaction of gene and environmental factors such as infectious agents, drugs or ionising radiations, etc.

ANALYSIS OF GENETIC DISORDERS

Family History

While dealing with a genetic case, the first step is recording the family history of the index case/proband. Proband is an affected person who has brought the family to the attention of a clinician. Proband is also called propositus (if male) or proposita (if female). The procedure starts with gathering information of the disorder, age at onset, duration of complaints and any other major illness. The second step is to collect information regarding the first-degree relatives, i.e. parents, siblings and offsprings of the proband. The following information is to be recorded—name, surname, date of birth, age, age and cause of death and relevant description of the disease if any. However, the following questions should help in the initiation of a dialogue with the patient or his relatives:

1. Does any relative suffer from similar trait?
   This will help in deciding the pattern of inheritance and subsequently working out the recurrence risk of the disorder.
2. Does any relative show any other disease that is not present in proband?
   For example, in case of dissecting aneurysm caused by Marfan syndrome, kindly ask about cardiac anomalies, ocular malformation or skeletal abnormality in relatives.
3. Ask for any condition with which any of the relatives has suffered or is suffering.
   Such a condition might have been unnoticed. For example, a propositus of pheochromocytoma can be suspected of having von Recklinghausen disease if patient's brother has scoliosis and mental retardation, since both are manifestations of this disease.
4. Is the proband an outcome of consanguineous marriage?
   Special attention should be given to this, because it (consanguinity) may lead to an autosomal recessive trait.
5. What is the ethnic group of the family?
   This is important because certain traits are common in some ethnic groups. African blacks show haemoglobinopathies such as HbS, HbC, β-thalassaemia and conditions like G6PD deficiency with much greater frequency.

The following points need to be emphasised while recording history:

1. Infant deaths, stillbirths and abortions to be noted with time of abortion, any obvious deformity in foetus or stillborn baby or in deceased infant. This may significantly alter the risk to subsequent pregnancy.
2. Illegitimacy should be borne in mind and proper enquiry with the family doctor or medicosocial worker of that area (in smaller place) may reveal facts.
3. "Nothing on my side" from anyone, during history taking, should arouse suspicion and should be verified with care.

Table 6.1 Symbols for Pedigree Charting

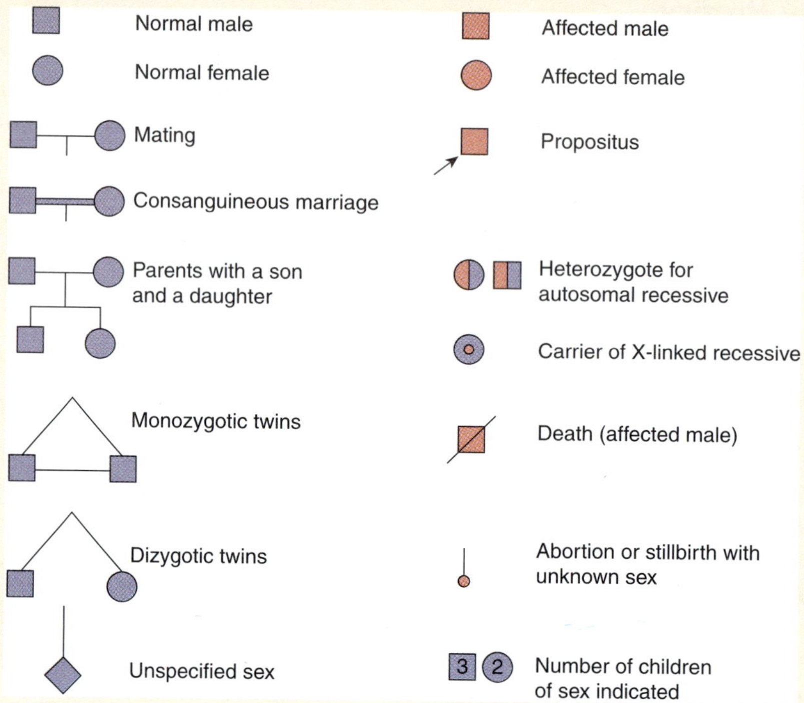

4. Record addresses of relevant family members as this is very important in order to contact them during follow-up whenever it is required.

Pedigree

It depicts the family data. It is a shorthand method of giving relevant information and also the mode of transmission of the disorder in the family. There are standard symbols used in drawing a pedigree. Table 6.1 shows symbols used in pedigree charting. In pedigree, the position of proband is shown by an arrow.

MENDELIAN INHERITANCE (SINGLE GENE DISORDERS) AN 74.1

They are caused by a single mutant gene. They follow one of the following four patterns of inheritance:

1. Autosomal dominant inheritance
2. Autosomal recessive inheritance
3. Sex/X-linked dominant inheritance
4. X-linked recessive inheritance

Usually, sex-linked inheritance means X-linked one because Y chromosome does not show any Mendelian genes except H-Y antigen genes. This forms the only trait transmitted from father to sons.

Terms "dominant and recessive" are used for the sake of convenience of pedigree analysis. These terms do not differentiate the genetic mechanism. When we say "dominant", it means that gene expresses clinically even in heterozygote state, i.e. in single dose. Similarly, the term "recessive" implies that double dose, i.e. homozygosity is essential for clinical manifestation of the trait. Kindly note that genes are never dominant or recessive. It is their effect, i.e. the trait/disorder, which is dominant or recessive.

Autosomal Dominant Inheritance

AN 74.1, 74.2

An autosomal dominant trait expresses in a heterozygote state. Homozygotes of it are severely affected because of a double dose of abnormal gene. It is often possible to trace an autosomal dominant trait through many generations. In fact, in South Africa, numerous cases of porphyria variegata have been traced back to a couple in 17th century. Porphyria variegata is metabolic disorder presenting skin blisters owing to increased sensitivity to sunlight. The urine becomes "port wine" coloured on standing due to porphyrins that it contains.

Pedigree analysis

AN 74.1, 74.2

In case of an autosomal dominant trait, pedigree shows the following criteria (Fig. 6.1):

1. An affected person has an affected parent, exception being mutant gene.
2. An affected person has normal and abnormal offsprings in equal proportion, i.e. there is 50% chance of dominant trait being transmitted to offsprings from affected parent.

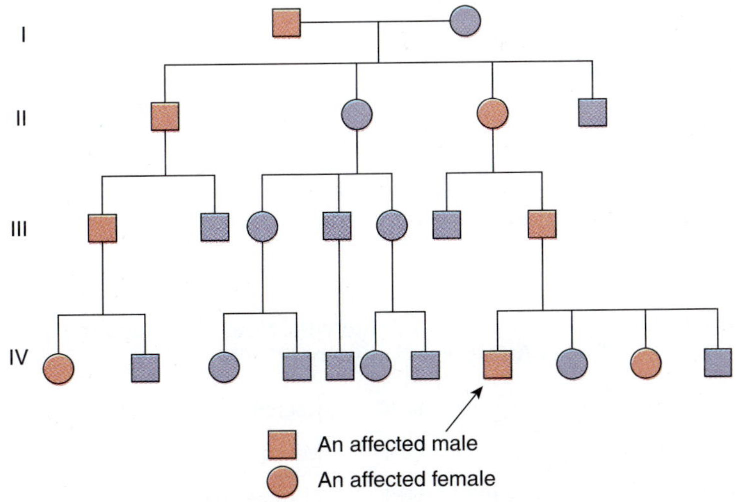

Figure 6.1 Pedigree of an autosomal dominant disorder.

3. Both males and females are equally affected.
4. The trait appears in every generation without skipping. An exception to this could be the trait impairing reproductive capacity of an affected person.
5. Normal children of an affected person do not transmit the disease.

Unaffected parent in dominant trait

AN 74.1, 74.2

In a dominant trait, usually parent is affected, but a normal parent can be expected under the following three conditions:

1. If the trait occurs because of a mutant gene (Table 6.2).
   In general, the frequency of mutation is 5×10^{-6} mutations per gene per generation. For frequency of mutation giving rise to an autosomal dominant trait, it works out to be 1 in 100,000 births. The dominant mutations usually involve gene coding for two classes of proteins:
   (a) Regulatory proteins, e.g. membrane receptors.
   (b) Structural proteins, e.g. collagen, haemoglobin.
2. Gene, though present in the parent, has low expressivity and so the parent appears to be normal (variability in expression of gene is described later in this chapter).
3. Another reason could be extramarital paternity. It accounts for about 3%–5% in the United States.

Two features associated with autosomal dominant disorder are as follows:

1. Delayed onset, e.g. Huntington chorea or adult polycystic kidney. Both the conditions manifest in adult life, although the mutant gene responsible for the trait is present at birth.
2. Variable clinical expression, e.g. multiple endocrine adenoma peptic ulcer syndrome. Here persons in the same family carrying the same abnormal gene may show hyperplasia of different endocrine glands giving rise to a varied clinical picture. There may be an involvement of pancreas, parathyroid, pituitary, etc. Accordingly manifestations in these individuals are different, e.g. hypoglycaemia, peptic ulcer, multiple lipomas of skin, kidney stones, bitemporal hemianopsia, etc.

Now let us consider the way transmission occurs in a dominant trait from parents to offsprings. Let us assume "A" as a mutant (dominant) gene and the normal allele being represented by letter "a" (Fig. 6.2).

An affected heterozygous parent with another normal partner shall have a progeny with a 50% chance of being affected and a 50% chance of being

Table 6.2 Percentage of Patients with New Mutations in Some Autosomal Dominant Traits

Achondroplasia	80%	Neurofibromatosis	40%
Tuberous sclerosis	80%	Marfan syndrome	30%
Treacher Collins syndrome	60%	Myotonic dystrophy	25%

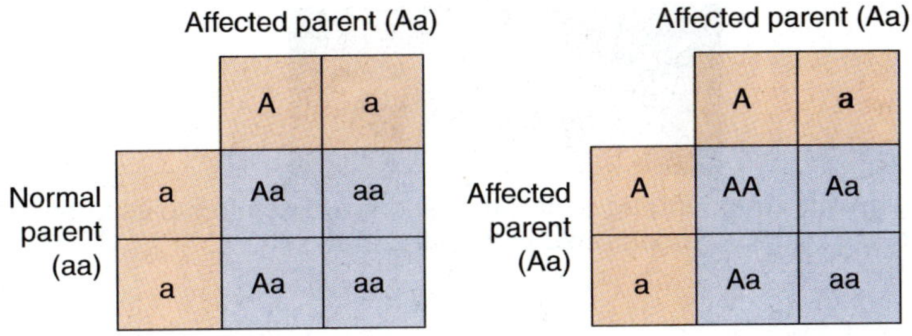

Aa = An affected heterozygote (50%)
aa = Normal homozygote (50%)

AA = An affected homozygote (25%) (severely affected)
Aa = An affected heterozygote (50%)
aa = Normal homozygote (25%)

Figure 6.2 Genetic risk in an autosomal dominant disorder.

normal. Let us consider the result of mating between the heterozygotes. This would result in a homozygous affected child. However, two persons both having the same rare dominant trait, chances are rare. All children of such homozygous person should have the trait.

Huntington disease (HD)

AN 74.1

Huntington disease claims the distinction of being the first genetic disease mapped to a specific chromosome with the help of RFLP marker. In 1983, J Gusella and his colleagues mapped the disease gene to short arm of chromosome 4. Nearly 10 years later this gene was cloned.

Huntington disease occurs with frequency of 1 in 20,000 Europeans and is less frequent in other populations. It is an autosomal dominant trait. It usually presents in age group of 30–50 years. It is characterised by progressive loss of motor control and dementia. Patients have difficulty in swallowing. Aspiration pneumonia, cardiorespiratory failure and subdural haematoma (due to trauma) are frequent causes of death in these patients. Choreic movements are often present in these patients.

Magnetic resonance imaging (MRI) shows substantial loss of neurons. There is decreased glucose uptake in brain. This forms an early sign of the disorder and can be detected by positron- emission tomography (PET). Though it involves many parts of the brain, it especially affects corpus striatum. Antipsychotic and antidepressant drugs help in controlling symptoms to large extent. Total clinical course of Huntington disease (HD) from diagnosis to death is about 15 years. Juvenile Huntington disease refers to onset of HD before 20 years and 80% of them are paternally transmitted.

Waardenburg syndrome

AN 74.1

It is an autosomal dominant trait. It may appear as a de novo event. Advanced paternal age has been observed as a factor in these fresh mutation cases. In cases, the mutations involved Pax3 gene located at 2q35. Mutations may involve other genes also, such as MITF gene at

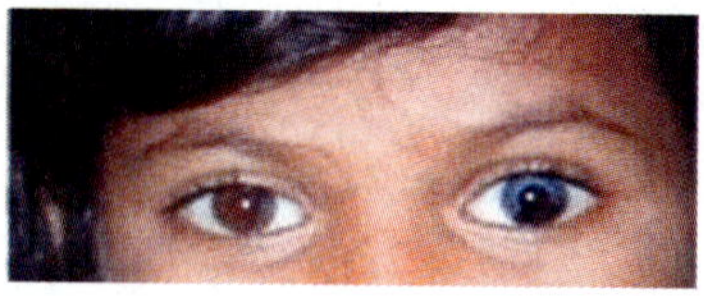

Figure 6.3 A girl with Waardenburg syndrome, showing lateral displacement of inner canthi, broad nasal bridge, hypochromic iridis with coloboma.

3p 12.3–p 14.1 or EDNRP or EDN3 or SOX10. The incidence is 1 in 42,000 births. Clinical features include lateral displacement of inner canthi with a broad and high nasal bridge, medial flare of eyebrows, partial albinism, white forelock, hypochromic iridis, deafness, etc. (Fig. 6.3).

Achondroplasia AN 74.4

It is an autosomal dominant disorder. A homozygous offspring shows severe skeletal deformities and usually dies during infancy. A heterozygote has a short stature, large head size, prominent forehead and scooped out nasal bridge. The limbs are short (Fig. 6.4). Achondroplastics usually have normal intelligence. Marriages between two achondroplastics are also known. About 80% cases are the result of new mutations.

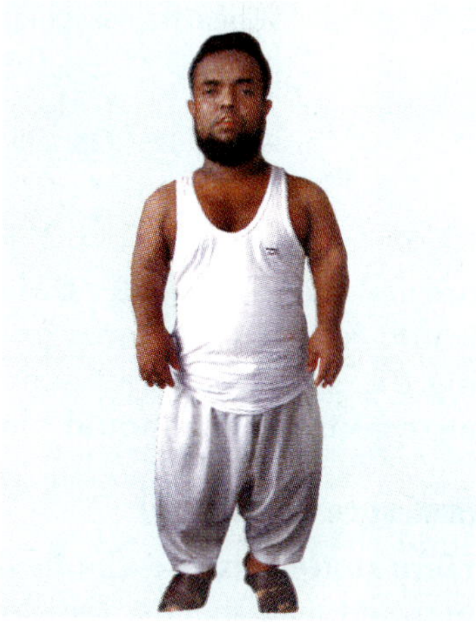

Figure 6.4 A patient of achondroplasia with short limbs. (*Modified:* This figure was published in Publication *Short Cases in Clinical Medicine*, 4th edition by A.B.M. Abdullah, ISBN: 9788131226698, page 209, Chapter 5, Copyright Elsevier, 2009.)

Tuberous sclerosis AN 74.1

It is an autosomal dominant trait manifesting in number of systems in variety of ways. It is an example of pleiotropy. It presents with a classical facial rash known as adenoma sebaceum, which may be confused with acne; however, microscopically it is angiokeratomas. It consists of blood vessels and the fibrous tissue. The patient also has learning difficulties and suffers from epilepsy. Nearly 80% cases are due to new mutations.

Treacher Collins syndrome AN 74.1

It is also called mandibulofacial dysostosis. The condition is named after Treacher Collins who described two cases of this disorder in 1900. The condition manifests with anti-mongoloid slant of palpebral fissures, malar hypoplasia and mandibular hypoplasia associated lower eyelid defects. Ear abnormalities involve external, middle and internal ear (Fig. 6.5). Defects in the latter two account for deafness. Cleft palate may also be present. Some patients show congenital heart disease, cryptorchidism and mental deficiency. The condition

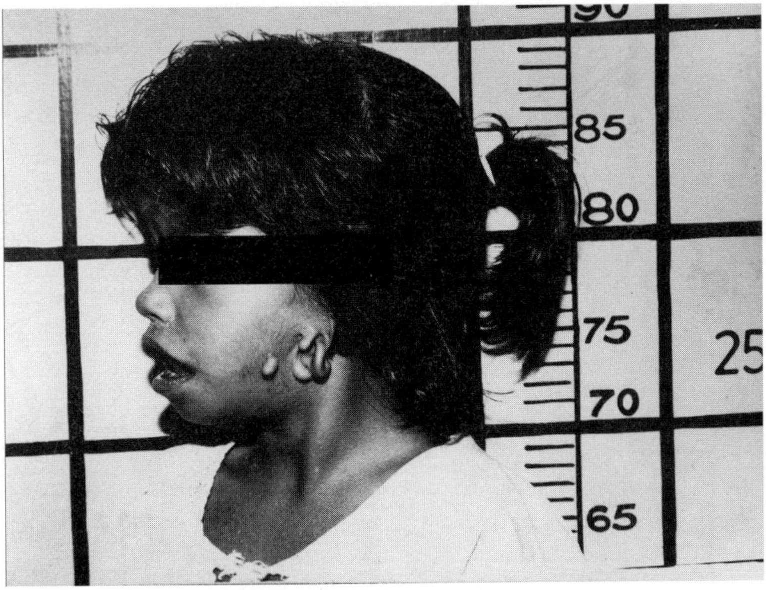

Figure 6.5 A girl with Treacher Collins syndrome.

follows autosomal dominant inheritance. In about 60% cases, there is fresh mutation. A recent study of affected families shows nearly 100% penetrance.

At this stage, I would like to draw your attention to two terms, *codominance* and *intermediate inheritance.* When both alleles of a pair are fully expressed in heterozygote, they are called *codominant.* For example, take an individual with AB blood group. He has both A and B antigens on his red blood cells. The allelic genes A and B are, therefore, codominant. Let us now consider another situation. In sickle cell anaemia, a heterozygote for abnormal allele does not have severe symptoms as found in homozygote. The heterozygote, however, possesses both, haemoglobin S as well as normal haemoglobin. In other words, a heterozygote shows intermediate expression between abnormal homozygote and normal homozygote. It is called sickle cell trait.

Progeria syndrome AN 74.1

Progeria syndrome is a rare genetic condition, characterised by the rapid, dramatic appearance of ageing beginning in childhood. Its incidence is 1 in 4 million newborns worldwide. This condition was first described by Hutchinson and Gilford in 1886, hence it is also called Hutchinson–Gilford progeria syndrome.

Clinical Features (Figs 6.6 and 6.7)

Affected children are apparently normal at birth, but after infancy growth is retarded and there is failure to thrive. The facial characteristics are typical with prominent eyes, a thin beaked tip nose, thin lips, a small chin and protruding ears. There is alopecia, mature sclerotic skin, joint contractures, bone abnormalities and subcutaneous loss of fat. The intellectual and motor skills development such as sitting, standing and walking are normal.

Figure 6.6 A boy with progeria syndrome, showing mature sclerotic skin, joint contractures. (Courtesy: Dr Shabana Borate, Associate professor, Grant Govt. Medical College, Mumbai.)

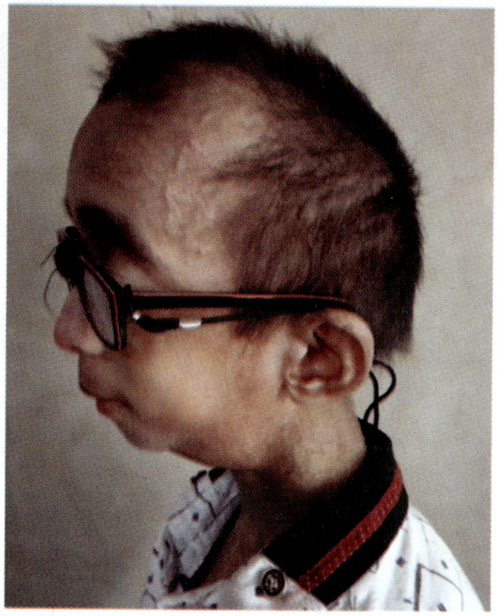

Figure 6.7 A boy with progeria syndrome, showing baldness, thin lips and prominent ear. (Courtesy: Dr Shabana Borate, Associate professor, Grant Govt. Medical College, Mumbai.)

There is prolonged prothrombin time and elevated platelet counts and serum phosphorus levels. There is measured reduction in range of joint motion, low-frequency conductive hearing loss and functional oral deficits.

The arteriosclerosis begins in the childhood. This condition greatly increases the chances of a heart attack or stroke at a young age. Death usually occurs around 13 years of age due to myocardial infarction or stroke.

Molecular Genetics

The genetic basis of this syndrome is a change from glycine GGC to glycine GGT in codon 608 of the lamin A (*LMNA*) gene. It activates a cryptic splice donor site to produce an abnormal lamin A which disrupts the nuclear membrane and alters transcription.

Margaret Casey was believed to be the oldest survivor of the premature ageing disease, who died at the age of 29-years.

Neurofibromatosis (Fig. 6.8) **AN 74.1**

Neurofibromatosis is an autosomal dominant trait, presenting mutation in about 40% cases. In neurofibromatosis type I, there is remarkable variability in the number of neurofibromas. Mental retardation may or may not be there. Hypertension may be found in various affected family members. It is due to stable mutation owing to insertion of one or a few base pairs. Diagnosis can be done on clinical examination.

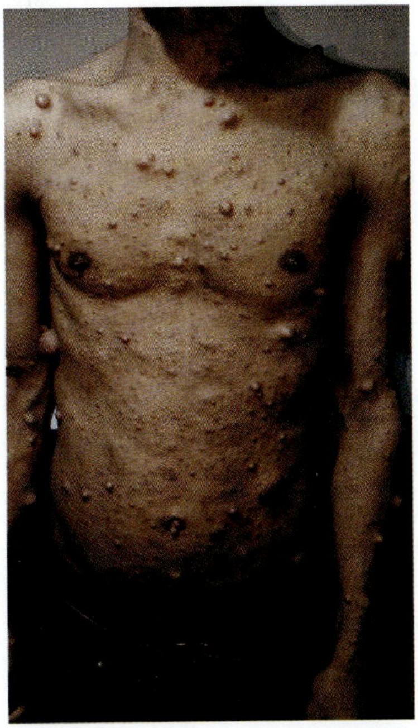

Figure 6.8 A case of neurofibromatosis, showing multiple nodules on trunk. (*Courtesy:* Dr Asha Kushwaha and Dr Pranit Farande, Dermatology Department, Terna Medical College, Navi Mumbai, Maharashtra, India.)

Neurofibromatosis type II may present with acoustic neuromas. Gene responsible for it is NF2. It may be associated with meningioma or gliomas. NF2 gene links cell surface molecule to the cytoskeleton. A careful clinical examination and investigation of the parents and related family members of the proband is an integral part in reaching the diagnosis.

Autosomal Recessive Inheritance

AN 74.1, 74.2

The recessive trait is expressed only in homozygote state. The homozygote receives one abnormal (recessive) gene from each parent. The trait typically appears only in sibs. The pedigree analysis of an autosomal recessive trait presents following features:

Features of autosomal recessive trait (Fig. 6.9)

AN 74.1, 74.2

These are as follows:

1. The trait appears in sibs and not in parents or offsprings.
2. About 25% of the sibs of the proband are affected, i.e. the risk is 1 in 4.
3. Both males and females have an equal chance of getting affected.
4. Parents of proband may be consanguineous.

A heterozygote for an autosomal recessive trait is called carrier. The word carrier has different connotation in medicine and genetics. In medicine, carrier is an individual harbouring an infective agent without clinical

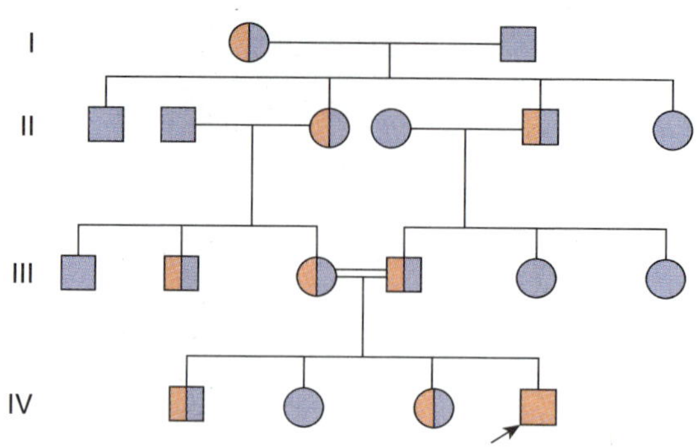

Figure 6.9 Pedigree of autosomal recessive trait.

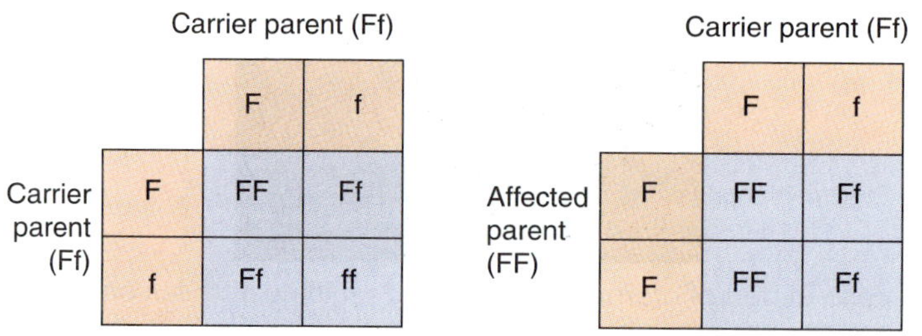

ff = Normal homozygote (25%)
Ff = Heterozygote carrier (50%)
FF = An affected heterozygote (25%)

FF = Affected (50%)
Ff = Heterozygote carrier (50%)

Figure 6.10 Genetic risk in an autosomal recessive disorder.

manifestation of the disease. In genetics, carrier means an individual in heterozygote state for gene determining inherited disorder who is essentially healthy at the time of examination. Meticulous clinical examination may reveal some features in favour of the disease. Alternatively, biochemical examination often reveals carrier state.

Now let us see situations in which parental carrier state reflects over progeny:

1. If both parents are carriers and both are heterozygote bearing one recessive gene "F" and one normal gene "f", mating result is—25% normal, 50% heterozygote carriers and 25% homozygote affected children (Fig. 6.10)
2. If one parent is heterozygote carrier and the other is homozygous affected, the progeny is as follows (Fig. 6.10): half the children will be affected and rest 50% will be heterozygote carriers. In this situation, the pedigree mimics an autosomal dominant inheritance.
3. If both parents are affected, i.e. they are homozygous for the abnormal gene, all the children will be (homozygous) affected.

Many inborn errors of metabolism follow this type of inheritance. This includes phenylketonuria, mucopolysaccharide disorders, such as Hurler syndrome and Tay–Sachs disease. Among common autosomal recessive entities comes cystic fibrosis, sickle cell anaemia; and rarer ones include Laurence-Moon-Biedl syndrome, Seckel syndrome, etc.

Cystic fibrosis (CF)

AN 74.4

It was recognised in 1936 as a separate clinical entity. It is the most common autosomal recessive disorder in Caucasians. Earlier it was called *mucoviscidosis* because of thick, viscid, mucous secretions that accumulate, block airways and cause secondary infections. Recurrent respiratory infection is mostly the cause of death.

Incidence

In Caucasians, incidence is 1 in 2000/3000. It is rarer in other populations as 1 in 17,000 in North Americans.

Life Expectancy

Till 1955, it was 5 years; now it is about 25 years.

Clinical Presentation

It presents with chronic lung disease, recurrent respiratory infection being the cause. Damage to lung leads to back pressure on right ventricle. This leads to cardiac failure. In almost 85% of patients with CF, pancreatic function is adversely affected. Pancreatic ducts are blocked (viscid secretions). Pancreatic enzyme secretion is reduced, leading to malabsorption. This also increases fat content of the stools.

Infertility in males is due to blockage of vas deferens. Other features include meconium ileus (blockage of small bowel), cirrhotic liver, rectal prolapse and so on.

Diagnosis

1. Elevated levels of sodium and chloride in sweat.
2. Measurement of immunoreactive trypsin (IRT) level in blood is elevated because of blockage of pancreatic duct.
3. Gene mapping of CF locus. The CF locus is on the chromosome 7q31 as found by linkage analysis.

The gene is known as Cystic Fibrosis Transmembrane Conductance Regulator (CFTR) gene. Its protein product is called CFTR. This protein is involved in the chloride transport and mucin secretion through the cell membrane. The most common mutation in *CFTR* gene is caused by deletion of the 508th codon. It is symbolised as Δ508.

Genetic Isolates

In some groups, one finds the frequent occurrence of otherwise rare recessive disorders. For example, the frequency of Tay–Sachs disease in general population is 1 in 360,000. However in the Ashkenazi Jews, the frequency of this disorder is 1 in 3600. In other words, the frequency of Tay–Sachs disease in Ashkenazi Jews is 100 times higher than general population.

This disease is characterised by neurological degeneration and cherry red spots in the fundus of eye. The age of onset is around 6 months. Affected children lack an enzyme, hexosaminidase A. Another example of rare recessive trait in genetic isolate is tyrosinaemia. It is a lethal hepatic disorder found in French–Canadian children of Quebec. Carrier frequency is 1 in 30 in this genetic isolate.

X-Linked Inheritance

AN 74.1, 74.2

Sex chromosomes are different in males and females. Male chromosome complement shows X and Y sex chromosomes, while female has two X chromosomes. This means when we talk about sex-linked inheritance, it can be either X-linked or Y-linked inheritance. Genes on Y chromosome show holandric inheritance. The Y chromosome bears H-Y antigen gene, which is of significance. The only recognisable trait of hairy pinna is associated with Y chromosome (Fig. 6.11). In short, this virtually means sex-linked inheritance is synonymous with X-linked inheritance. Under X-linked inheritance, it can be either recessive or dominant. As such X-linked dominant traits are relatively rare, so let us now consider X-linked recessive inheritance.

Males are said to be hemizygous in respect of X-linked gene since they have only one X chromosome. An X-linked recessive trait is determined by a gene carried on the X chromosome. It manifests in females only when it is in double dose (homozygous state). Therefore, females are rarely affected. A heterozygous female forms a carrier. In males who are hemizygous for X-linked genes, the trait expresses in single dose. This means males with single mutant gene on their X-chromosome are affected.

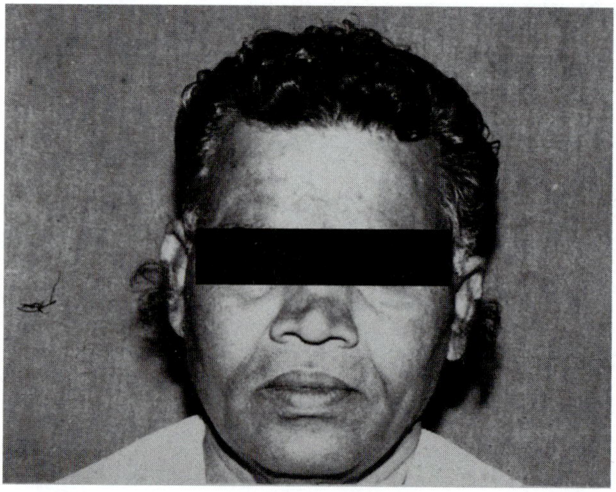

Figure 6.11 Hairy pinna, a trait associated with Y chromosome.

		Mother carrier (X_HY_h)	
		X_H	X_h
Normal father (X_HY)	X_H	X_HX_H	X_HX_h
	Y	X_HY	X_hY

(X_HX_h) = Carrier daughter, X_hY = An affected son

Figure 6.12 Checkerboard showing offsprings of carrier mother and normal father in case of haemophilia.

		Mother (X_HX_H)	
		X_H	X_H
Father affected (X_hY)	X_h	X_HX_h	X_HX_h
	Y	X_HY	X_HY

All sons are normal and all daughters are carriers

Figure 6.13 Checkerboard showing offsprings of an affected male and normal female in haemophilia.

There are numerous conditions under X-recessive traits, e.g. haemophilia, Duchenne muscular dystrophy, colour blindness, G6PD deficiency, etc. Let us consider the progeny in case of haemophilia.

1. Mother X_HX_h and father X_HY.
2. Mother X_HX_H and father X_hY. Here "H" stands for dominant allele and "h" for recessive gene for haemophilia on X chromosome.

The outcome of this mating in the first situation (Fig. 6.12) is that half of the daughters are normal, while the other half are carriers of the trait. Similarly, half of the sons are normal and the other half are affected.

In the second situation (Fig. 6.13), all the daughters are heterozygote carriers and all the sons are normal.

Features of X-linked recessive inheritance (Fig. 6.14) **AN 74.1, 74.2**

Features of X-linked inheritance are as follows:

1. The trait affects males (rarely females).
2. The trait is transmitted from an affected male through all his daughters to half of their sons.

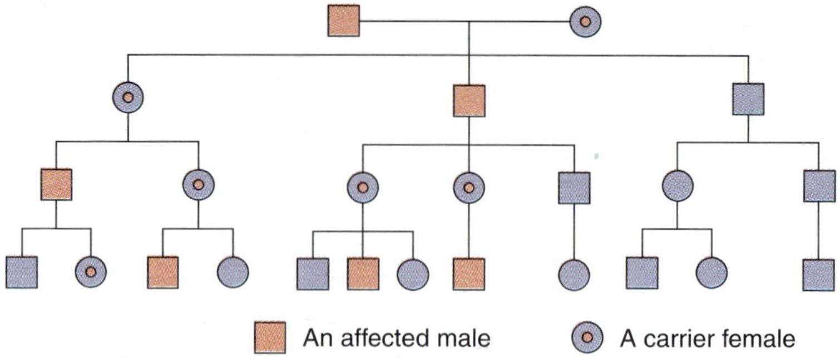

Figure 6.14 Pedigree of X-linked recessive disorder.

3. No male to male transmission occurs, since father contributes only Y chromosome to his sons.
4. In a kindred, affected males are related to each other through carrier females.

Haemophilia (Figs 6.12 and 6.13) AN 74.4

It is an X-linked recessive trait. The incidence of the disorder is 1 in 10,000 male births. In this, males are affected and females are carriers. The chance of a female getting affected is remote. The basic defect is that these individuals have a deficiency of factor VIII (anti-haemophilic globulin) in the blood. Because of this deficiency, the blood does not clot. On injury, these persons bleed profusely. Clinical manifestation includes severe arthritis because of haemorrhages within the joints. This particular X-linked disorder caught attention because Queen Victoria was a carrier of the disorder.

Duchenne muscular dystrophy (DMD) AN 74.4

It is an X-linked recessive trait, characterised by progressive muscular weakness. It is named after a French neurologist Guillaume Duchenne who described it in 1861. Mutations involving the same gene lead to a milder condition called Becker Muscular Dystrophy (BMD).

Duchenne

(*Attributed to*: St. George's University of London.)

Incidence

It is 1 in 3500 males, while BMD is encountered with the frequency of 1 in 20,000.

Clinical Features

Mostly DMD boys present between 3 and 5 years. They have muscular weakness manifesting as an awkward gait, difficulty in climbing and inability to run. A typical problem is found when the child wants to rise from the floor. The child attempts it by pushing on or "*climbing up*" his legs and thighs. It is called *Gower sign*. Delay in learning to walk around 18 months in DMD boys (normal child does it around 12 months) usually brings them for consultation.

Lumbar lordosis is seen in advance cases. Increasing difficulty in walking confines them to wheelchair by the age of 11 years in most cases. It is followed by joint contracture, respiratory failure and rarely cardiac failure. The mean age about which death occurs is 18 years. IQ, mean value is 83. The intellectual impairment is as a result of pleiotropic effect of DMD gene mutation. BMD presents in milder form with an average age of onset, 11 years. Patients remain ambulant till adult life.

Diagnosis

1. Creatine kinase level in serum is elevated; in normal boys—200 i.u.; in DMD boys between 1000 and 20,000 i.u.
2. Muscle biopsy shows increased variation in fibre size in early stages, and necrosis followed fibrous replacement in subsequent stages.

DMD locus and Gene: The DMD locus has been mapped to Xp 21 by linkage analysis. In two-third cases, one can identify gene deletion. The protein product of *DMD* gene is called *dystrophin. DMD* gene is the largest gene so far identified in humans. It consists of 2300 kilobases of the genomic DNA. It has 79 exons. Its site of expression is muscle; in some patients, it is in neurons of the cerebral cortex.

Hot Spots: They are the deletion sites in *DMD* gene detected after Southern blot analysis with cDNA probes. The first is located in the first 20 exons of the gene, while the other is in the centre of the gene between 45 and 53 exons.

Management

There is no effective and satisfactory treatment available today; however, several approaches have been tried. These are:

1. Direct injection of recombinant DNA.
2. Myoblast implantation.
3. Transfection with retroviral or adenoviral vectors containing a *dystrophin minigene.*

X-linked Dominant Inheritance

AN 74.1, 74.2, 74.4

In contrast to the X-linked recessive trait, the X-linked dominant trait occurs more frequently in females. It is twice as common in females as in males. The affected male transmits the trait to all his daughters and none to his sons. This distinguishes X-linked dominant inheritance from an autosomal dominant inheritance. Pedigrees in both types of dominant inheritance otherwise resemble closely. Tracing progeny of an affected female does not offer a clue in distinguishing between an autosomal and an X-linked dominant inheritance. Examples of X-linked dominant inheritance are: Xg blood group, vitamin D-resistant rickets, hypophosphataemia, etc.

Features of X-linked dominant inheritance (Fig. 6.15)

1. Trait is more frequent in females than in males who present relatively milder expression of the disorder.
2. Affected male transmits the trait to all his daughters and not to his sons.
3. Affected homozygote female transmits trait to all children.
4. An affected heterozygote female transmits the trait to half her children of either sex. This mimics autosomal dominant transmission; hence, to distinguish between autosomal and X-linked dominant inheritance, one has to follow progeny of affected male.

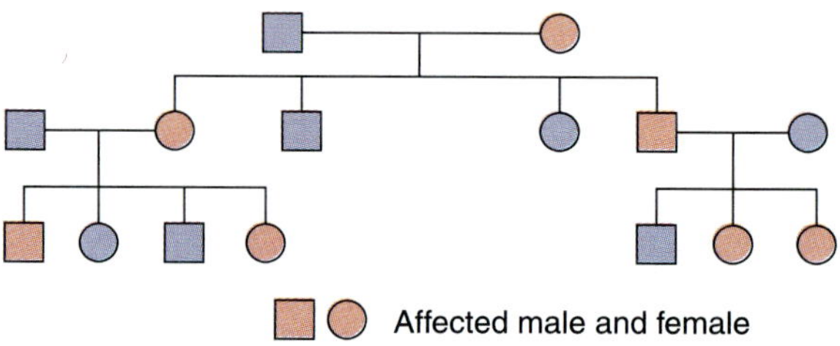

Figure 6.15 Pedigree of X-linked dominant disorder.

Fragile X

AN 74.1

At this stage, it is worth mentioning a unique feature in medical genetics called "Fragile X syndrome". It is unique in the sense that it is caused by a combination of a mutant gene with an associated cytogenetic abnormality. It is clear that the mutant gene is X-linked, but it is difficult to say whether it is dominant or recessive.

Clinical presentation of a fragile X case includes large prominent ears, large-sized testes after puberty along with mental retardation. In fact, fragile X syndrome forms one of the important causes of mental retardation in males. Cytogenetic study needs little modification in the procedure of karyotyping. The culture medium used to grow cells is deficient in folic acid and thymidine. Low concentration of these substances in a medium gives better results. Affected males show fragile X in about 35% cells. Carrier females may not show fragile X at all. In some cases even if they show fragile X, it is with low frequency. It is often located at the end of the long arm of X chromosome (X_q 27–28). It was observed for the first time by Lubs in the USA. It may be noted that fragile sites do occur along other chromosomes apart from X chromosome, but their significance is unknown to date.

Gene: Expression and Penetrance

Wide diversity in the expression of a gene causing a disorder may lead to problems in correct diagnosis and interpretation of pedigree, especially in an autosomal dominant trait. Expressivity of a gene refers to a degree of expression, and in clinical terms it can be mild, moderate or severe form of the disease. The word penetrance implies whether the gene will be expressed or not. If a person has the mutant gene responsible for the trait, but it fails to express, the trait is said to show reduced penetrance. In fact, penetrance follows what can be called all-or-none law. If the trait caused by the same gene shows different forms/severity in different members, it is said to exhibit variable expressivity. Clinical severity of the disorder in different members shows variety of phenotypic effects.

Pleiotropy

Normally, each gene has one primary effect, i.e. it directs synthesis of a polypeptide. However, when a single gene or a gene pair produces multiple phenotypic effects, it is called "pleiotropy". There are plenty of examples among various clinical syndromes under this category. In an autosomal recessive disorder called phenylketonuria (PKU), an enzyme phenylalanine hydroxylase is deficient. This enzyme deficiency is a primary defect. It leads to multiple secondary effects. They are severe mental retardation, PKU (passing phenyl ketones in urine), hypopigmentation, etc. Galactosaemia forms another example of pleiotropy. It is characterised by lack of an enzyme galactose-1-phosphate-uridyl-transferase. Secondary effects of the enzyme deficiency being cirrhosis of liver, cataract, mental retardation and galactosuria. In such events, the secondary untoward effects can be prevented by deletion of the specific component from diet, such as phenylalanine-free diet be given to a baby with PKU. This will prevent mental retardation.

In some syndromes, signs and symptoms are related to a single structural defect. For example, in osteogenesis imperfecta type 1, the basic defect is in collagen synthesis. This accounts for multiple secondary effects like brittle bones, osteosclerosis, blue sclerae, etc. In yet another condition called Marfan syndrome, the primary defect lies in synthesis of elastic fibres. This exhibits in pleiotropic manifestations such as skeletal, ocular and cardiovascular anomalies.

Genetic Heterogeneity

This is in contrast to pleiotropy. In genetic heterogeneity, several genes produce one effect. On clinical examination, the traits appear to be indistinguishable. However, each trait on genetic analysis reveals a different picture. Each trait could be because of mutation at different sites or a different type of mutation at the same locus, for example, deafness. In 1976, Fraser studied this condition. According to him, there are about 16–18 types of autosomal recessive deafness. Under autosomal dominant type, we have Waardenburg syndrome, in which deafness is associated with white forelock, pale irides. Another autosomal dominant condition showing deafness as a feature is first arch syndrome. Likewise, there are X-linked conditions with associated deafness. In short, deafness is a common presenting clinical manifestation in all of them. Each condition, however, arises from an involvement of genes at different loci or different mutation at the same locus. This is called "genetic heterogeneity". There are several examples of this; they include muscular dystrophies and osteogenesis imperfecta, the latter having four major types.

Sex-Limited Traits

We realise that if a trait is determined by an X-linked gene, then it will not affect males and females in equal proportion, i.e. the sex ratio will not be 1:1. In other words, if the sex ratio deviates from 1:1, it indicates an X-linked disorder. Such a deviation, however, is not only associated with X-linked but also with autosomal genes. The traits that are determined by autosomal transmission but are expressed only in one sex are called *sex-limited traits.*

One of them is precocious puberty. This affects males. They are heterozygous. These individuals develop secondary sexual characters and an increment in height much early, around 4–5 years. As compared to normal boys of their age, they are taller in the initial period. Early fusion between epiphyses and diaphyses, however, makes them finally short men. The trait shows an autosomal dominant inheritance.

Sex-Influenced Traits

The trait is said to be sex-influenced when it expresses in both males and females but with different frequencies. For example, baldness is expressed in males as an autosomal dominant trait. Congenital adrenal hyperplasia is more commonly expressed in females.

Multifactorial Inheritance AN 74.3

A multifactorial trait is defined as one that results from a combination of factors, genetic as well as non-genetic, each exhibiting a minor role. It is believed to account for much of the normal variations seen in families, and numerous common disorders too.

The term polygenic inheritance is commonly used alternatively with multifactorial inheritance. This is because it is often difficult to decide whether environmental factors are operational in bringing about the defect or whether all the genes determining the trait have small additive effects. The family patterns of many normal traits and genetic defects go in favour of multifactorial inheritance.

Among normal traits, we have family patterns peculiar of multifactorial inheritance. For example, height or stature of an individual, intelligence quotient (IQ) and total ridge count (TRC). All three traits are measurable, i.e. stature in centimetres of height, TRC as number of ridges, and IQ as

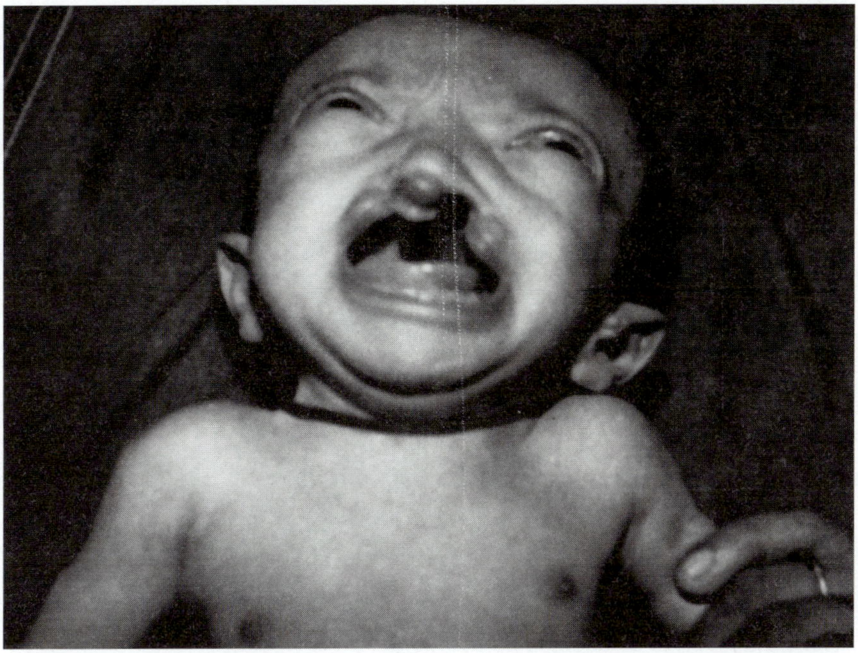

Figure 6.16 A child with cleft lip and cleft palate (multifactorial inheritance).

points. Both genetic as well as environmental factors appear to be instrumental in governing all the three traits that are mentioned above.

Some of the common congenital malformations show multifactorial inheritance. For example, cleft lip, cleft palate (Fig. 6.16), club foot, congenital dislocation of hip joint (CDH), congenital heart disease (CHD), and neural tube defects such as anencephaly and spina bifida, etc. Non-genetic factors operational under multifactorial inheritance include teratogenic agents. They are:

1. Drugs, e.g. thalidomide, anti-convulsants, anti-cancer drugs, anti-malarials, etc.
2. Infections, e.g. rubella virus.
3. Ionising radiations, e.g. X-rays and administration of radioactive substances such as I^{131}, P^{32} and Au^{198}.

Summary

The genetic disorders are classified into three categories: (*i*) chromosomal disorders; (*ii*) single gene disorders and (*iii*) multifactorial inheritance.

Analysis of Genetic Disorder: It involves the following steps:

Family history: Does any relative suffer from similar trait? Does any relative show any other disease? Ask for any condition in past (gone unnoticed). Is proband an outcome of consanguinity? What is ethnic group?

Continued

Summary—cont'd

Record of the following:

i) Infant deaths, stillbirth and abortions.
ii) Illegitimacy is to be borne in mind.
iii) Address and contact numbers of family members.

Pedigree: It depicts the family data and gives information about mode of inheritance.

Mendelian Inheritance (Single Gene Disorders)

Caused by single gene mutation; they exhibit four patterns:

i) Autosomal dominant
ii) Autosomal recessive
iii) Sex/X-linked dominant
iv) X-linked recessive inheritance

Autosomal Dominant Inheritance: An autosomal dominant (AD) trait is expressed in heterozygote state. Homozygotes are severely affected.

Pedigree Analysis

a. An affected person has affected parent.
b. An affected person has normal and abnormal offsprings in equal proportion
c. Both males and females are equally affected.
d. Trait appears in every generation.
e. Normal children of an affected person do not transmit disease.

Unaffected Parent in Dominant Trait

a. If trait is due to mutant gene, e.g. achondroplasia, tuberous sclerosis.
b. Gene present in the parent has low expressivity.
c. Extramarital paternity.

An AD trait shows two features: (*i*) Delayed onset, e.g. Huntington chorea and (*ii*) Variable clinical expression, e.g. multiple endocrine–adenoma–peptic ulcer syndrome.

Features of Autosomal Recessive (AR) Trait: Expressed in homozygous state only.

i) The trait appears in sibs and not in parents.
ii) About 25% sibs of proband are affected.
iii) Both males and females are equally affected.
iv) Parents of the proband may be consanguineous.

If both parents are carriers, the children shall be 25% normal, 50% heterozygote carriers and 25% homozygote affected. Many inborn errors of metabolism follow AR inheritance, e.g. PKU, mucopolysaccharidosis, cystic fibrosis, etc.

Cystic Fibrosis (CF): Occurs as a common AR trait in Caucasians. It presents with recurrent respiratory infection, pancreatic dysfunction, malabsorption and infertility in males. Diagnosis is possible with (*i*) elevated Na and Cl levels in sweat; (*ii*) raised immunoreactive trypsin (IRT) in blood; and (*iii*) gene mapping. CF locus is at 7q31. The gene is Cystic Fibrosis Transmembrane Conductance Regulator (CFTR) gene.

Genetic Isolates: A group that has frequent occurrence of otherwise rare disease, e.g. Tay–Sachs disease in Ashkenazi Jews. The disease frequency is 100 times more than the general population. Other example is tyrosinaemia found in French–Canadian children of Quebec.

X-linked Inheritance: H-Y antigen gene is the only significant gene, and hairy pinna as the only recognisable trait associated with Y chromosome. In short, sex-linked inheritance is synonymous with X-linked inheritance.

Summary—cont'd

Features of X-linked Recessive Inheritance:

a. Trait affects males (rarely females).
b. Trait is transmitted from affected males through his carrier daughters to half of their sons.
c. No male-to-male transmission occurs.
d. In kindred, affected males are related through carrier females.

Haemophilia: An X-linked recessive trait affecting males. It occurs with frequency of 1 in 10,000 male births. Basic defect is deficiency of factor VIII (anti-haemophilic globulin) in blood. Blood does not clot. An injury causes profuse bleeding and haemorrhage in joints leads to severe arthritis. Queen Victoria was carrier of the disorder.

Duchenne Muscular Dystrophy (DMD): An X-linked recessive trait characterised by progressive muscular weakness, named after a French neurologist Guillaume Duchenne. *Incidence* is 1 in 3500 male births. *Features* are muscular weakness, awkward gait, Gower sign, lumbar lordosis, and later on joint contractures, respiratory failure and cardiac failure. *Diagnosis*: Creatinine kinase level is elevated, muscular biopsy reveals increase in fibre size, later on necrosis and fibrous replacement. *DMD Locus and Gene:* Locus is at Xp21 and the protein product of DMD gene is dystrophin. DMD gene is the largest gene in man with 2300 kb size. There are deletion sites in DMD gene, detected on Southern Blot. *Management*: Direct injection of recombinant DNA, myoblast implantation and transfection with retroviral vectors containing dystrophin minigene.

X-Linked Dominant Inheritance: It is more frequent in females. An affected male transmits the trait to all his daughters and none to his sons, e.g. vitamin D resistance rickets, hypophosphataemia.

Pedigree Features:

- Trait is more frequent in females.
- Affected male transmits the trait to all his daughters and not to his sons.
- Affected female, if homozygote, transmits trait to all her children.

Fragile X: This syndrome is unique; it is caused by combination of mutant gene with an associated cytogenetic abnormality. *Clinical features* include prominent ears, large testes, mental retardation. *Karyotype* reveals fragile X in 35% of cells at Xq 27–28.

Gene Expression and Penetrance: Expressivity of gene refers to the degree of expression; it can be mild, moderate or severe. The word penetrance implies whether the gene will be expressed or not (follows all-or-none law).

Pleiotropy: It refers to multiple phenotypic effects caused by single gene or gene pair, e.g. Phenylketonuria. It is an AR trait with deficiency of phenylalanine hydroxylase enzyme as primary defect. It leads to multiple secondary defects such as mental retardation, phenylketonuria, hypopigmentation, etc. In some syndromes, e.g. osteogenesis imperfecta type I, the basic defect is in collagen synthesis; this leads to multiple secondary effects such as brittle bones, blue sclerae and so on.

Genetic Heterogeneity: It is in contrast to pleiotropy, i.e. several genes produce one effect. Clinically the traits are indistinguishable. However, each trait on analysis reveals mutation at different sites or different types of mutation at the same locus, e.g. deafness. Fraser has demonstrated 16–18 types of autosomal recessive deafness.

Continued

Summary—cont'd

Sex-Limited Traits: The traits that are determined by autosomal transmission but are expressed in one sex only, e.g. precocious puberty in males.

Sex-Influenced Traits: Trait is so called when it expresses in both males as well as females but with different frequencies, e.g. baldness is expressed in males, congenital adrenal hyperplasia in females.

Multifactorial Inheritance: Trait that results from combination of both factors, genetic as well as non-genetic, e.g. height/stature, intelligence quotient (IQ), total ridge count (TRC). Some common congenital malformations are—cleft lip, cleft palate, congenital dislocation of hip, etc.

QUESTION YOURSELF*

1. What are the criteria of an autosomal dominant trait?
2. How will you explain an unaffected parent in dominant trait?
3. All of the following are autosomal dominant traits except:
   a. Tuberous sclerosis
   b. Duchenne Muscular Dystrophy
   c. Treacher Collins syndrome
   d. Neurofibromatosis
4. What are the features of pedigree of an autosomal recessive trait?
5. What is the connotation of the word "carrier" in genetics?
6. Which one of the following mode of inheritance is found in cystic fibrosis?
   a. Autosomal recessive
   b. Autosomal dominant
   c. X-linked recessive
   d. Multifactorial
7. Which one of the following is the site of CF locus?
   a. 13q14
   b. 11p21
   c. 7q31
   d. Xq22
8. What does CFTR stand for?
9. What is genetic isolate?
10. Why sons of an affected haemophilic father are normal?
11. Duchenne muscular dystrophy is transmitted as:
    a. Autosomal recessive trait
    b. Autosomal dominant trait
    c. X-linked recessive trait
    d. X-linked dominant trait
12. What is peculiar of DMD gene?
13. What are "hot spots"?
14. What is the unique feature of fragile X syndrome?
15. What is pleiotropy?
16. What is meant by sex-limited trait?
17. What is meant by sex-influenced trait?

18. Name four traits with multifactorial inheritance.
19. What is pedigree?
20. Match the following:

Clinical Entity	**Mode of Inheritance**
1. Cystic fibrosis	a. Autosomal dominant
2. Hypophosphataemia	b. Autosomal recessive
3. Tuberous sclerosis	c. X-linked dominant
4. Duchenne muscular dystrophy	d. X-linked recessive

*See pages 261–262 for Answers.

7 Biochemical Genetics

COMPETENCY AN 74.4 Describe the genetic basis and clinical features of sickle cell anaemia.

LEARNING OBJECTIVES

At the end of this chapter the students should be able to understand

- Basis of inborn errors of metabolism
- Haemoglobinopathies and types of haemoglobin
- Genesis of thalassaemias

KEY WORDS

Phenylketonuria (PKU), Mucopolysaccharidosis (MPS), G6PD deficiency

Biochemical events in the living organism whirl around DNA and proteins encoded by it. The proteins make either structural protein or enzymes that govern various metabolic processes. The proteins synthesised in the body influence growth as well as differentiation. Any alteration in DNA information (i.e. gene mutation) will result in the production of variant protein, and this in turn has phenotypic effect. Such an effect may involve an amino acid substitution, altering structural protein such as haemoglobin. The phenotypic effect of it may be in the form of its altered affinity for oxygen or tendency to sickle. An alteration in amino acid sequence in an enzyme may alter its activity, usually decreasing it, but occasionally this alteration in enzymic structure may enhance its activity. Genetically determined variations also manifest in altered response to drugs in some individuals. In short, let us now define the extent and scope of biochemical genetics and also what we shall discuss under this chapter:

1. Gene mutations leading to inborn errors of metabolism.
2. Haemoglobinopathies and normal haemoglobin.
3. Polymorphisms revealed by an altered response to drugs.

INBORN ERRORS OF METABOLISM

Grouped under this heading are the biochemical disorders in which there is an enzyme defect (Table 7.1). This defect is genetically determined. It produces alteration in the metabolic process. The first example of this was cited

Table 7.1 Few Important Inborn Errors of Metabolism

Disorder	Enzyme	Inheritance	Clinical Manifestations
Acatalasia	Catalase	AR	May present as oral gangrene
Albinism	Tyrosinase	AR may be X-linked	Lack of pigment in skin, hair and eyes (Fig. 7.1)
Alkaptonuria	Homogentisic acid oxidase	AR	Arthritis
Cystinuria	Not known	AR	Renal stones, aminoacid-uria
Galactosaemia	Galactose-1-phosphate uridylyltransferase	AR	Mental retardation, cataract, cirrhosis
Gaucher disease	Glucocerebrosidase	AR	Hepatosplenomegaly, thrombocytopenia and anaemia
G6PD deficiency	Glucose-6-phosphate dehydrogenase	X-linked	Haemolysis in response to some drugs
Hunter syndrome	Sulphoiduronate sulphatase	XR	Hepatosplenomegaly, mental retardation, skeletal abnormalities
Hurler syndrome	α-Iduronidase	AR	Same as above, in addition there is corneal clouding
Isoniazid slow inactivator	N-acetyltransferase	AR	Neurological problems
Niemann–Pick disease	Sphingomyelinase	AR	CNS damage, cherry red spot on macula and hepatosplenomegaly
Phenylketonuria	Phenylalanine hydroxylase	AR	Microcephaly, mental retardation
Porphyria (acute intermittent)	More hepatic ALA synthetase	AR	Acute abdominal pain episode, neurological problems, excessive excretion of amino-levulinic acid (ALA) in urine
Tay–Sachs disease	Hexosaminidase-A	AR	Convulsions, mental retardation, blindness
Vitamin-D resistant rickets	Renal defect in phosphate reabsorption	XD	Rickets
Wilson disease	Not known	AR	Cirrhosis of liver, Kayser–Fleischer ring in cornea, neurological problems

(Adapted and Modified from: Emery's Elements of Medical Genetics, Eleventh Edition, Table 10.1, Page 152, Churchill Livingstone.)

Figure 7.1 A case of albinism showing lack of pigment in skin, hair and eyes. (*Picture Courtesy:* Dr. Asra Khumushi, Dermatologist, Cosmetologists and Laser Surgeon, Nagpur.)

by Garrod in 1902, as alkaptonuria. He noted that the excretion of alcapton or homogentisic acid in these patients is because of the failure of breakdown of benzene ring of tyrosine. However, it took long time for the first specific enzyme defect to be demonstrated leading to an inborn error. In 1952, Gerty Cori demonstrated that lack of glucose-6-phosphatase leads to Von Gierke disease.

Garrod

(*Source: Metabolism: Clinical and Experimental:* Archibald Edward Garrod: the physician father of biochemistry; Anna Piro, Antonio Tagarelli, Giuseppe Tagarelli, Paolo Lagonia, Aldo Quattrone. Elsevier, April 2009.)

Genesis of Inborn Errors of Metabolism

Alteration in the DNA sequence is the basic defect responsible for inborn errors of metabolism. It causes change in the structure of an enzyme or protein coded by it. This defective enzyme shall have:

1. reduced activity because its affinity towards substrate is reduced, or
2. the structurally different enzyme molecule may be unstable with resultant error in metabolism.

Pathogenesis in these disorders is explained as under:

1. Accumulation of precursor due to lack of enzyme activity, e.g. phenylketonuria.

2. The end product of metabolism being less, manifests accordingly, e.g. congenital adrenal hyperplasia.

In non-metabolic errors, the genesis of disease could be different, e.g. familial hypercholesterolaemia. In this, the cell surface receptors are defective for low density lipoprotein (LDL). As a result, LDL-bound cholesterol remains outside the cell. In yet another non-metabolic error called I-cell disease, the enzymes do not reach lysosomes resulting in lysosomal malfunction.

Phenylketonuria (PKU)

The specific enzyme defect in this disorder was demonstrated by Jervis in 1953. The defective enzyme found was phenylalanine hydroxylase. Steps in phenyl-alanine metabolism and the related disorders are shown in Fig. 7.2.

In PKU, lack of phenylalanine hydroxylase causes phenylalanine to follow an alternative pathway. This converts it into phenylpyruvic acid and other toxic metabolites. These are excreted in the urine.

Manifestation

A child born with PKU is normal at birth, but subsequently as it receives phenyl-alanine in diet, toxic metabolites of phenylalanine metabolism accumulate causing damage. This leads to mental retardation.

In addition, phenylalanine is not converted into tyrosine. So this reduces proportion of melanin. This results in PKU child presenting with blond hair, blue eyes and lack of pigment in the brain, e.g. in substantia nigra.

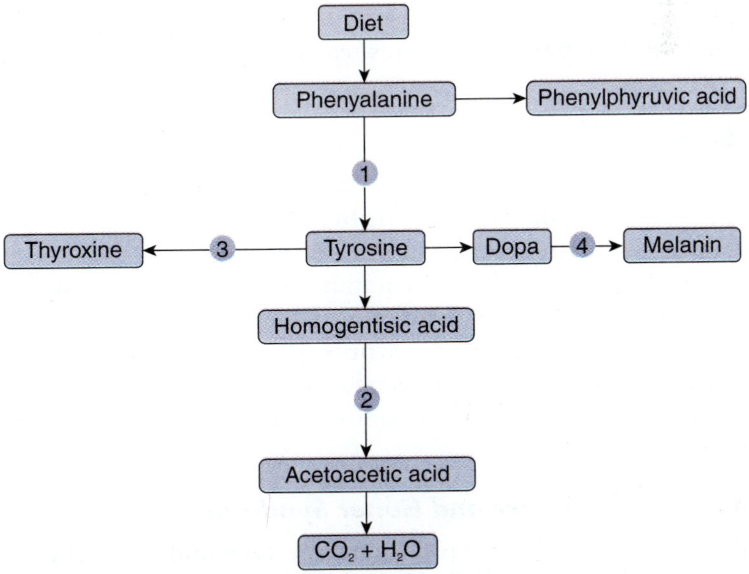

Figure 7.2 Phenylalanine metabolism in body. The related disorders are: (1) phenylketonuria, (2) alkaptonuria, (3) dyshormonogenesis and (4) albinism.

Diagnosis

It can be accomplished by:

1. Ferric chloride test: Detection of phenylpyruvic acid in urine.
2. Guthrie test: It is bacterial inhibition assay to detect excess of serum phenylalanine.

Treatment

The ideal way to treat these children would be to offer them defective enzyme. Somehow it is not possible and hence we resort to other alternatives, i.e. to eliminate phenylalanine from the diet. Since it forms one of the essential amino acids, it cannot be removed from the diet completely. So, what one can do is to give a controlled amount of phenylalanine in the diet, simultaneously monitoring the blood level of this amino acid. Early detection of the defect is essential, otherwise mental retardation results. Amount of damage once caused is irreversible.

Mucopolysaccharidoses (MPS)

It is a group of lysosomal storage disorders. It results from defective degradation of carbohydrate side chain of acid mucopolysaccharides. It causes accumulation of glycosaminoglycans, which leads to problems in the central nervous system, vascular system or skeletal system. There are about a dozen conditions described under this group. Of these, two claim their position in the following descriptions.

Hunter syndrome

It was first described in 1917. It is inherited as an X-linked disorder. The deficient enzyme is sulphoiduronate sulphatase. There is accumulation of dermatan and heparan sulphates causing multiple problems in the body.

Hurler syndrome

It was first demonstrated in 1919. It follows autosomal recessive inheritance. The deficient enzyme responsible for this condition is α-iduronidase. There is defective degradation of mucopolysaccharides causing accumulation of heparan and dermatan sulphates. It presents a similar clinical picture as in Hunter syndrome except that corneal clouding is found in Hurler syndrome. Hurler syndrome is usually fatal in childhood. It is believed that transfusion of plasma or leucocytes may serve as a source of deficient enzyme in these patients. This may show a temporary improvement.

Clinical Features of Hunter and Hurler Syndromes

The clinical features include typical gargoyle face and short stature associated with skeletal malformations. The children are mentally retarded. The disorder is progressive and is manifested as severe or mild form.

HAEMOGLOBINS AND HAEMOGLOBINOPATHIES

Geneticists owe a lot to haemoglobins, which have offered a molecular basis of human genetics. Haemoglobin is a transporter of oxygen. There is a group of disorders of haemoglobins called haemoglobinopathies. To understand them, it becomes essential for us to study the structure, function and formation of haemoglobin.

Structure of Haemoglobin

It is a protein in red blood cells. It is engaged in the transport of oxygen. The molecular weight is 64,500 (daltons). The molecule has two components, the first one is the "haem" part responsible for oxygen transport. It is similar in all types of haemoglobins. The other component is the "globin" portion. It is this part that varies in various forms of haemoglobins.

Haemoglobin molecule has its globin portion formed by four polypeptide chains. In HbA (haemoglobin in adult), there are two α chains and two β chains (non-α chains). HbA is expressed by the formula $\alpha_2\beta_2$. Both α and β chains are almost equal in length. The α chain has 141 amino acids and β chain has 146 amino acids—a total of 287 amino acids. α chain is coded by α gene on chromosome 16; β chain is coded by β gene on chromosome 11. Since they are located on different chromosomes, mutation may involve either α chain or β chain, but not both. The haemoglobin molecule shows eight helical regions. The iron atom haem is attached to histidine by a bond, the only link between haem and globin portions. Other non-α chains are δ, γ and ϵ chains. They are more like β chains. They are present in various forms of haemoglobins. Table 7.2 shows a few of the abnormal forms found in human beings.

Table 7.2 Some Normal and Abnormal Forms of Haemoglobins

Haemoglobin	Structure	Percentage in Adult
HbA	$\alpha_2\ \beta_2$	97–98
HbA2	$\alpha_2\ \beta_2$	2–3
HbF	$\alpha_2\gamma_2$	Less than 1
Gower I	$\zeta_2\epsilon_2$	
Gower II	$\alpha_2\epsilon_2$	
Abnormal forms		
HbS	$\alpha_2\ \beta_2\ (\beta_2{}^6$ glu–val)	
HbC	$\alpha_2\ \beta_2\ (\beta_2{}^6$ glu–lys)	
HbH	β_4	

(*Adapted and Modified from: Emery's Elements of Medical Genetics,* Eleventh Edition, Table 9.1, Page 140, Churchill Livingstone.)

Animation 3: Haemoglobin structure. (*Available online at* www.medenact.com)

Let us consider some clinically important abnormal presentations of haemoglobins.

Haemoglobin S: Sickle Cell Disease

AN 74.4

It was the first abnormal haemoglobin identified. In 1949, Pauling identified physico-chemical alterations in haemoglobin S by electrophoresis. On further analysis, it was found that β globin chain of HbS is different from HbA. Valine replaces glutamic acid in the sixth position of β chain in HbS molecule. This change is responsible for the difference in electrophoretic mobility of HbS and HbA. In the heterozygote of sickle cell, there is a mixture of HbS and HbA, while homozygote has only HbS type of haemoglobin. Heterozygote presents as sickle cell trait, a milder form that appears to be clinically normal. Homozygote presents as sickle cell disease. These patients have red cells with a tendency to sickle. This leads to anaemia and subsequently splenomegaly and weakness. The red cells tend to clump/cluster. This in turn causes thrombosis/infarction and ischaemia. The total effect is in the form of abdominal pain, splenic infarction, bony tenderness, haematuria, renal failure and heart failure.

Haemoglobin C

AN 74.4

It shows a replacement along β chain very much like HbS. The replacement involves glutamic acid replaced by lysine in the sixth position of HbA from N-terminal. This form of haemoglobin is found in Africa.

Lepore haemoglobin

AN 74.4

In this type, the haemoglobin has normal α chain but non-α chain has a part homologous to N-terminal of δ chain and a part homologous to C-terminal of normal β chain. Together they form δβ chain. This happens because δ and β genes (coding for these chains) resemble to 136 sites out of 146. It means, they differ only at 10 sites. The genes coding for these chains are altered. During meiosis because of mismatching, there is formation of a new gene with portions of both δ and β genes. This is a fusion product and is called *Lepore gene,* associated with deletion of part of each locus.

The fusion product could be with β gene portion followed by δ gene portion. This is called "anti-Lepore gene". It is associated with duplication (Fig. 7.3).

Thalassaemias

The word thalassaemia is derived from a Greek word *Thalassa* meaning sea. It is also called Mediterranean anaemia according to its distribution. This forms a group of disorders with problems of haemoglobin synthesis. The structure of haemoglobin is not defective. Let us see what is the problem in synthesis. Haemoglobin as we know consists of α and β chains. Now imagine a situation in which the rate of synthesis of any one of these chains is

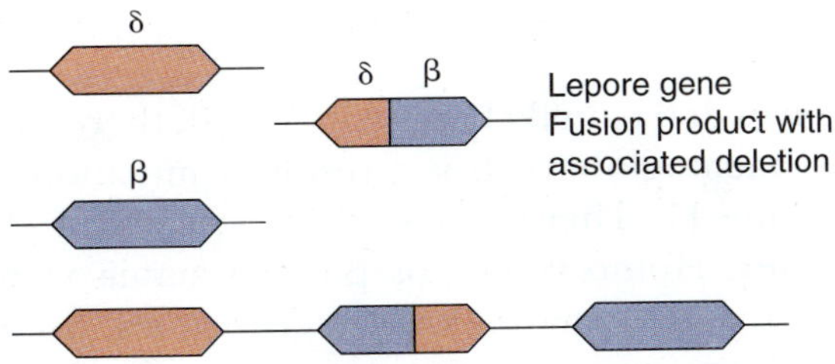

Figure 7.3 Diagrammatic representation of formation of Lepore and anti-Lepore genes.

reduced. This means if α chain synthesis rate is lowered, there will be excess of β chains (comparatively) having normal synthesis rate. This creates problems with maturation and survival of erythrocytes. There are two main groups:

α-Thalassaemias

They are characterised by a lowered rate of synthesis or an absence of synthesis of α chains. This naturally affects both foetal as well as adult haemoglobin synthesis. To understand the defect, let us go back to the gene level. The α chains are coded by α genes. There are two of them with each chromosome 16. This means, in total there are four α genes, each accounting for 25% of α globin component. What is the effect of these genes? In a homozygote of α-thalassaemia, all four genes are absent, so no α chain synthesis occurs. In short, if all four genes are present a total of 100% synthesis of αchain occurs; with two genes it is 50%, and with one gene it is 25%. The deletion/mutation of a progressively higher number of genes accounts for more severe abnormality (Table 7.3).

Table 7.3 Correlation Between Genotype and a Chain Synthesis

Genotype	Synthesis of α Chain
—/—	0%
α –/—	25%
α –/α or α α/—	50%
α α/α –	75%
α α/α α	100%

(*Adapted and Modified from:* Thompson and Thompson, *Genetics in Medicine*, Sixth Edition, Table 11.3, Page 195, Saunders, 2004.)

β-Thalassaemias

The disorder involves defect in β chain synthesis. It, therefore, affects synthesis of HbA (adult haemoglobin). It is a result of mutation or deletion of β gene on chromosome 11. There is one locus on each 11 chromosome, i.e. there are two genes. Homozygotes of β-thalassaemia with both β genes altered present as thalassaemia major. Heterozygotes, however, present in milder form. The homozygotes have severe anaemia. β-Thalassaemia may result from deletion of a part of β gene complex or it may be an outcome of Hb Lepore, i.e. δβ fusion gene.

An attempt was made to treat patients of β-thalassaemias by using 5-azacytidine. It promotes γ globin synthesis. This increases synthesis of HbF ($\alpha_2\gamma_2$). The drug is also helpful in sickle cell anaemia. Prenatal diagnosis of homozygous β-thalassaemia can be done by foetal blood sampling around the 18th week of gestation. This helps in genetic counselling and management of the disorder.

PHARMACOGENETICS AND DISORDERS RELATED TO DRUGS

I am not very good at mathematics; however, the equations placed below allow me to think. Have a look at them.

	Patient		Drug		Result	
1	😐	+	℞	=	☺	Cured
2	😐	+	℞	=	☹	No relief
3	😐	+	℞	=	☠	Death

In the first *situation*, the patient suffering from a disease gets cured with the administration of drug. *In the second*, the same drug is being used for the same disease, but patient has no relief. *In the third situation*, the drug and the disease being the same but the result is disastrous, patient succumbs to the disease/dies.

In short, what we analyse from these situations is that in all the three cases, the disease and the drug were same; however, the results are much different. In one, we found complete cure; in the second, no relief; and in the third, the individual showed adverse reaction. Here the variable is the person, i.e. patients' genetic profile governs as to how patient will respond to a particular drug. This brings us to the term "Pharmacogenetics".

The term pharmacogenetics was first used by Vogel. It deals with genetically determined variations that become evident with an altered response to drugs. The scope of pharmacogenetics encompasses investigations to reveal the cause of:

1. Antibiotic resistance in some strains of bacteria.
2. Resistance to organophosphorus compounds (used as insecticides) in some insects. This has application in public health.

3. Defining potency of a drug.
4. Explaining sensitivity of some individuals to a particular drug.

Fate of Drug

A majority of drugs follow an underlying course in the body (Fig. 7.4). Transformation of drugs involves some important processes, viz.:

1. **Conjugation:** It occurs chiefly in the liver. It may be glucuronide conjugation. Morphine and codeine are processed in this manner.
2. **Acetylation:** It also occurs in the liver and involves an addition of acetyl group to the original molecule. Isoniazid, an anti-tubercular drug, follows this manner of inactivation. Sulphonamides also follow a similar pattern. We shall now see some of the polymorphisms for drug response.

Acatalasia

Persons with acatalasia do not have an enzyme catalase. This enzyme breaks down hydrogen peroxide into oxygen and water. It is present in the blood. This condition, acatalasia, was discovered by a Japanese clinician Takahara in 1946, while treating a girl for oral gangrene. After the removal of gangrenous tissue, he poured hydrogen peroxide to sterilise the wound. He observed that blood turned brownish black in colour on contact with hydrogen peroxide. He suggested that it was probably due to lack of catalase in red cells of the patient, and haemoglobin was oxidised to methaemoglobin thus giving a brownish–black colour. Under this condition, there are three types of individuals:

1. Person homozygous for normal gene has a normal enzyme level.
2. Person homozygous for acatalasia gene has no enzyme in the blood.
3. Person heterozygote has a moderate level of enzyme in the blood.

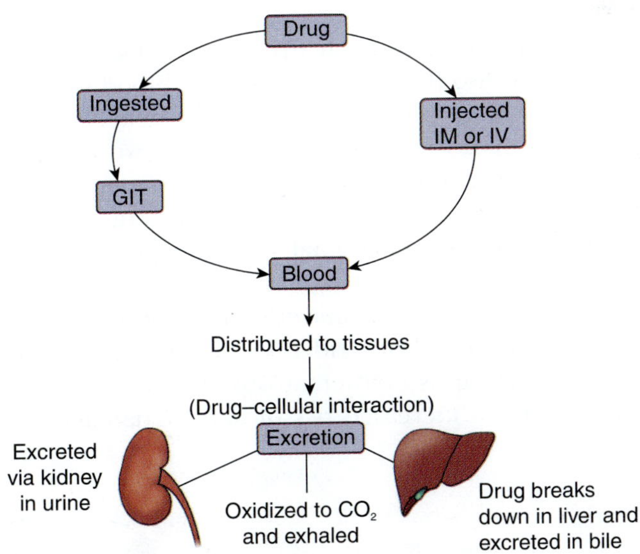

Figure 7.4 Drug metabolism in body.

Glucose-6-phosphate-dehydrogenase deficiency

It is inherited as an X-linked recessive disorder. In these individuals, there is a deficiency of red cell G6PD. They suffer from haemolytic episodes with primaquine, a drug used to prevent relapses in case of malaria. On administration of the drug, there are no adverse effects initially. Afterwards, the patient begins to pass dark/black urine, develops jaundice and his haemoglobin percentage reduces. These persons are sensitive to many drugs, e.g. phenacetin, sulphonamides, aspirin, Furadantin, etc.

Succinylcholine sensitivity

It is inherited as an autosomal recessive trait. Succinylcholine is a drug that is widely used as a muscle relaxant in anaesthesia. Its molecule structurally consists of two molecules of acetylcholine. It is rapidly hydrolysed by an enzyme serum cholinesterase. Previously it was called pseudocholinesterase, because its hydrolytic action on acetylcholine is slower than red cell enzyme "true cholinesterase". In patients having succinylcholine sensitivity, the enzyme does not destroy the drug at normal rate. This may cause prolonged apnoea in these patients. Now it is possible to study plasma pseudocholinesterase by using a local anaesthetic dibucaine. It is expressed as a dibucaine number.

Malignant hyperthermia

It may rarely follow as an anaesthetic complication. It forms a rare autosomal dominant trait. There is hyperpyrexia to the tune of 108°F, usually following the use of halothane as anaesthetic agent and succinylcholine as relaxant for intubation. The basic defect is reduced uptake and binding of calcium ions in sarcoplasmic reticulum.

Isoniazid activation

Isoniazid is used as an anti-tubercular drug. The drug is used orally. It gets absorbed from the gut into the blood, raising the blood isoniazid level. This is followed by inactivation and excretion of the drug, reducing its blood level. Considering isoniazid metabolism, there are two types of individuals, slow and rapid inactivators of isoniazid. The drug is metabolised by acetylation, and involves an enzyme, hepatic *N*-acetyltransferase. The slow inactivators are homozygous for an autosomal recessive gene for the hepatic *N*-acetyltransferase.

The clinical implication of the isoniazid inactivation study is that the slow inactivators may suffer from isoniazid toxicity. This may be manifested as polyneuritis or systemic lupus erythematosus (SLE)-like disorder. However, rapid inactivators carry an increased risk of hepatic damage.

Summary

Biochemical events in an organism whirl around DNA and proteins encoded by it. Alterations in DNA (i.e. mutation) lead to variant protein production with phenotypic effect.

Inborn Errors of Metabolism (IEM): A genetically determined enzyme defect alters metabolic process. In 1902, Garrod demonstrated the first example as alkaptonuria.

Genesis of IEM: An alteration in DNA sequence causes change in structure of an enzyme or protein. The defective enzyme may have reduced activity or may be unstable with resultant metabolic error. Pathogenesis of the disorder can be explained in two ways: (*i*) Accumulation of precursor, e.g. PKU; (*ii*) end product of metabolism being less has its effect, e.g. congenital adrenal hyperplasia.

Phenylketonuria (PKU): Demonstrated by Jervis in 1953, the disorder is due to deficiency of phenylalanine hydroxylase. This converts it into phenylpyruvic acid and other toxic metabolites. Accumulation of these toxic metabolites leads to mental retardation. Phenylalanine cannot be converted into tyrosine; this reduces production of melanin and hence leads to blond hair, blue eyes, lack of pigment in substantia nigra.

Diagnosis: (*i*) Ferric chloride test; (*ii*) Guthrie test.

Treatment: Give controlled amount of phenylalanine in the diet, simultaneously monitoring its blood level.

Mucopolysaccharidoses (MPS): It is a group of lysosomal storage disorders, with defective degradation of carbohydrate side chain of acid mucopolysaccharide. It leads to accumulation of glycosaminoglycans with problems in CNS, CVS and skeletal systems, e.g. *Hunter syndrome.* It is an X-linked disorder. Enzyme sulphoiduronate sulphatase is deficient, causing accumulation of dermatan and heparan sulphates. This presents with typical gargoyle face, short stature, skeletal malformations and mental retardation.

Hurler syndrome: It is an autosomal recessive trait. Enzyme α-iduronidase is deficient, causing defective degradation of mucopolysaccharides and accumulation of dermatan and heparan sulphates. Clinically, it has similar features as Hunter syndrome, but in addition there is corneal clouding.

Haemoglobins and Haemoglobinopathies

Structure of haemoglobin molecule: It has a molecular weight of 64,500 daltons. It consists of two parts, "haem" (O_2 transporter) and "globin". It is the globin part that varies in various forms of haemoglobins. Four polypeptide chains, two α and two β (non-α chain) make globin part. HbA (haemoglobin in adult) is expressed as $\alpha_2\beta_2$. α chain has 141 amino acids, and β chain has 146 amino acids. α chain is coded by α gene on chromosome 16, and β chain is coded with β gene on chromosome 11. Since they are located on different chromosomes, mutation may involve either one or both. The Hb molecule shows eight helical regions with iron atom haem attached to histidine by a bond, the only link between haem and globin fractions. Other non-α chains found in different forms of haemoglobins are δ, γ and ϵ chains.

Haemoglobin S—Sickle cell disease: Haemoglobin S (HbS) was the first abnormal haemoglobin found in 1949. In this, valine replaces glutamic acid in sixth position of β chain in HbS molecule. This alters its electrophoretic mobility. A homozygote has only HbS and suffers from sickle cell disease, while a heterozygote of sickle cell has a

Continued

Summary—cont'd

mixture of HbS and HbA and suffers from a milder form called sickle cell trait. In sickle cell disease/trait, the RBCs tend to clump/cluster, in turn causing thrombosis, infarction, abdominal pain, bony tenderness, haematuria, etc.

Haemoglobin C: It is found in Africa; in this, glutamic acid is replaced by lysine in sixth position of HbA.

Lepore haemoglobin: In this haemoglobin, α chain is normal but the non-α chain has a part homologous to N-terminal of δ chain and a part homologous to C-terminal of normal β chain. Together they form δβ chain. This is because δ and β gene (coding for these chains) resemble to 136 sites out of 146, thus differing only at 10 sites. A fusion product is δβ gene (Lepore gene). A fusion product with β gene portion followed by δ gene portion is called "anti-Lepore gene".

Thalassaemias

Thalassa, a Greek word means sea. It is also called Mediterranean anaemia owing to its distribution. It is of two types:

- **α-Thalassaemia:** It is characterised by lowered rate of synthesis or absence of synthesis of α chains of the haemoglobin molecule. This is due to deletion/mutation involving α gene/s on chromosome 16. There are total four α genes; depending on number of genes involved it will manifest in mild or severe forms.
- **β-Thalassaemia:** It involves β chain synthesis. There are two β genes. In homozygotes both genes are altered and they present as Thalassaemia major while heterozygote presents milder form of Thalassaemia. An attempt to treat β Thalassaemia with 5-azacytidine that promotes γ globin synthesis have been made.

Pharmacogenetics: It deals with genetically determined variations in response to drugs. The scope of pharmacogenetics includes antibiotic resistance in bacteria, resistance to organophosphorus compounds in insects, defining potency of drugs and explaining sensitivity of some persons to a particular drug.

Fate of a Drug: Transformation of a drug involves glucuronide conjugation or addition of an acetyl group, i.e. acetylation in liver.

Acatalasia: In this, enzyme catalase in RBCs is lacking, therefore haemoglobin is oxidised to methaemoglobin giving brownish-black colour on exposure. Both homo- and heterozygotes for acatalasia gene are known.

G6PD deficiency: It is an X-linked recessive disorder. The individuals with G6PD deficiency of red cells suffer from haemolytic episodes with primaquine, an antimalarial drug. They are sensitive to many drugs, e.g. phenacetin, sulphonamides, aspirin, etc.

Succinylcholine sensitivity: It is a muscle relaxant used in anaesthesia. It is hydrolysed by serum cholinesterase. In patients with succinylcholine sensitivity, the drug is not hydrolysed at normal rate and may result in prolonged apnoea.

Malignant hyperthermia (108°F): It is anaesthetic complication and is an AD trait. Halothane, an anaesthetic agent, and succinylcholine, a relaxant, used in intubation may lead to malignant hyperthermia in some individuals. This is because of reduced uptake and binding of Ca ions in sarcoplasmic reticulum.

Isoniazid activation: Isoniazid is an anti-TB drug. Its inactivation involves enzyme N-acetyltransferase. These are of two types. The slow inactivators are homozygous for an AR gene for N-acetyltransferase and may suffer from isoniazid toxicity, while rapid inactivators are at risk of hepatic damage.

QUESTION YOURSELF*

1. What is meant by inborn error of metabolism?
2. What is PKU?
3. What do you mean by haemoglobinopathy?
4. What is pharmacogenetics?
5. What is the scope of pharmacogenetics?
6. Who coined the term pharmacogenetics?
7. Enumerate the processes by which drugs are dealt within the body.
8. What is acatalasia?
9. What is the mode of inheritance in G6PD deficiency?

*See pages 262–263 for Answers.

8 Genetics of Blood Groups

COMPETENCY AN 75.4 **Describe the genetic basis of polymorphism.**

LEARNING OBJECTIVES

At the end of this chapter the students should be able to understand

- Basis of formation of blood groups and common blood group systems
- Other blood group systems

KEY WORDS

ABO blood group, Bombay phenotype, Haemolytic disease of newborn (HDN)

The first breakthrough in the history of blood group system was in the year 1900—Landsteiner discovered ABO blood group system this year. Study of the blood group system offers understanding of some of the genetic principles, such as multiple alleles, polymorphism, immune reactions, linkage phenomenon, etc.

Among various components of human blood, red cell antigen forms one of the genetic markers. This can be used as a genetic marker in population studies or in linkage analysis. This is possible, because it satisfies the following criteria:

1. It forms different phenotypes.
2. It follows a simple pattern of inheritance.
3. Its frequency is different in different populations.
4. It is not influenced by environmental factors or age.

Landsteiner

(*Source: History of forensic serology and molecular genetics in the sphere of activity of the German Society:* D. Patzelt, Fig. 2, Vol. 144: 2-3, Elsevier Ireland Ltd, 2004.)

CLINICAL APPLICATIONS

Clinically significant blood polymorphisms include ABO system, Rh blood group and HLA system. In general, the term blood group refers to red blood cell antigens. In the following conditions, they are important:

1. **Blood transfusion:** In this, ABO compatibility is checked by blood grouping and cross-matching.
2. **Tissue transplantation:** Here apart from ABO compatibility, HLA typing of both the donor and the recipient is done to rule out antigenic disparity and achieve graft survival.
3. **Haemolytic disease of newborn (HDN):** It is an outcome of Rh incompatibility between the mother and conceptus. This can be avoided by taking suitable measures (details of this are given later in this chapter).

ABO BLOOD GROUP SYSTEM

This came into existence through the discovery by Landsteiner in Vienna. According to this, there are two antigens on the red blood cells. They are antigen A and antigen B. Their presence or absence gives rise to four phenotypes; they are A, B, AB and O.

With the presence of an antigen, one can expect antibodies in the sera of these individuals. A person with blood group "A" has an antigen A on his red cells and antibody anti-B in his serum. An individual with "O" group has neither A nor B antigen on his red cells. So he possesses both antibodies, anti-A and anti-B in his serum.

The ABO system forms an example of multiple allelism. The alleles are A, B and O genes located at ABO locus on the long arm of chromosome 9. Please note that gene O is recessive to genes A and B, while A and B are codominants. If both A and B genes are present, then both antigens (A and B) are formed. The O gene is an amorph, i.e. it has no effect, thus leaving H-substance unaltered (Fig. 8.1).

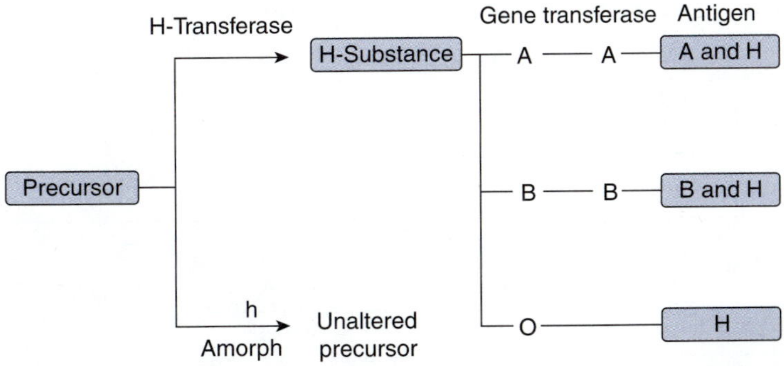

Figure 8.1 Biosynthesis of H, A and B antigens. Genes O and H do not have detectable effect.

The blood group antigen–antibody reaction is called isoagglutination. It is by this reaction that the blood group of an individual can be determined with the help of standard anti-sera that are available. Table 8.1 shows the possible agglutination reaction. It can be observed as a clumping of red cells. Theoretically, one can assume group O as a *universal donor* and group AB as a *universal recipient*. However, it is desired that transfusion is done of the same blood group.

So far, numerous subtypes of A and B have been detected. Significant among them are A_1 and A_2. As a result of this, the AB group is also subdivided into A_{1B} and A_{2B}. Almost 85% of blood group A consists of A_1. Other variants of A and B, although known, are rare.

Diseases and ABO System

Association between a particular disease and blood group was suggested first in 1953. It was observed that patients with gastric cancer show an excess of group A individuals. Another close association was observed between group O and duodenal ulcer. Type A individuals also show increased tendency to clotting, eventually raising the risk of thrombosis. This observation may prove to be significant in thrombosis occurring in group A women on oral contraceptives.

Bombay Phenotype or O_h Phenotype

This rare phenotype was first identified in Bombay (1952) and hence labelled as Bombay phenotype. In this rare phenotype, the red blood cells as well as secretions lack antigens A, B and H. The serum contains anti-A, anti-B and anti-H antibodies. Individuals with O_h phenotype are homozygous for gene h; and have genotype hh. They do not form H antigen, and hence no

Table 8.1 Agglutination Reaction in Various ABO Phenotypes

Phenotype	Genotype	Antigen on RBC	Antibody in Serum	Reaction with Blood Groups			
				A	B	AB	O
A	AA	A	Anti-B	–	+	+	–
	AO						
B	BB	B	Anti-A	+	–	+	–
	BO						
AB	AB	A and B	No	–	–	–	–
O	OO	No	Anti-A				
			Anti-B	+	+	+	–

"+" represents agglutination; "–" represents no agglutination.

substrate is available for making antigen. Therefore A and B genes, although present, cannot form the respective antigens.

Secretor Status and ABH Antigens

These antigens are found not only in red blood cells but in most other cells of the body. They are also present in secretions like saliva. The ability to secrete AB and H antigens is governed by "Se", i.e. secretor gene. The secretors have genotype as "SeSe" or "Sese", and account for about 78% of the population. Non-secretors have genotype "sese". Non-secretors do not have A, B or H antigens in their (secretions) saliva but have them on their red blood cells. The secretor gene locus is on long arm of chromosome 19. The steps involved in the synthesis of H, A and B antigens are shown in Fig. 8.1.

RHESUS BLOOD GROUP SYSTEM

It is named after the Rhesus monkey because it was used in experiments, which finally culminated in the discovery of this system. It was discovered in 1940 by Landsteiner and Wiener. The Rh system claims its place because of its role in HDN. There are both Rh-positive and Rh-negative individuals.

Rh-positive persons are homozygous or heterozygous for gene specifying an antigen D, while Rh-negative persons do not have antigen D. The Rh locus is on chromosome 1. Fisher and Race suggested that there are five Rh antigens designated as D, C, E, c, and e. There are eight alleles. Their antigenic determinants are shown in Table 8.2.

Rh-Null Blood Group

The persons of this blood group do not have Rh antigens on their red blood cells. This is comparable to Oh phenotype. Rh-null persons present with

Table 8.2 Alleles of the Rh Blood Group System

Allele	Antigen
R^0	D, c, e
R^1	D, C, e
R^2	D, c, E
R^z	D, C, E
r	c, e
r′	C, e
r″	c, E
r^y	C, E

haemolytic anaemia indicate that Rh antigens form an important and integral part of red cell membrane. In the absence of these antigens, red cells become vulnerable. These individuals lack precursor that is needed as a substrate for production of antigens of Rh system. Rh-null phenotype was first encountered in Australian aborigines. Rh-null individuals are homozygous for gene X^0r; its allele $X'r$ is needed for synthesis of Rh as well as LW antigens. Thus, Rh-null people also lack LW antigens.

Clinical Significance

Any Rh-negative person, on exposure to Rh-positive red blood cells, forms anti-Rh antibodies. Care is to be taken in Rh-negative girls and women of child-bearing age that they receive Rh-negative blood in transfusion therapy if the transfusion is required. Rh-negative pregnant women may be given Rh immune globulin injections during pregnancy and just after the parturition. This will minimise the risk of immunisation against Rh antigen in them.

Haemolytic Disease of Newborn (HDN) (Fig. 8.2)

In HDN, foetal red blood cells die earlier due to the action of antibodies formed by the mother against foetal Rh antigen. It begins in utero but continues after birth for about 3 months—the time taken by maternal antibodies to get cleared from newborn's circulation. Although it is based on genetically determined antigenic disparity between mother and the baby, it is an acquired haemolytic anaemia, different from hereditary ones like hereditary spherocytosis.

In the natural course, there are no antibodies against Rh antigen in serum. During pregnancy, foetal and maternal blood pools are isolated by placental barrier. Towards the term, however, there are breaks occurring along this barrier. This permits transfer of foetal red cells to the mother's blood. When the mother is Rh-negative and foetus is Rh-positive, this transfer of Rh-positive red cells of foetus shall evoke an antibody response in the mother. These

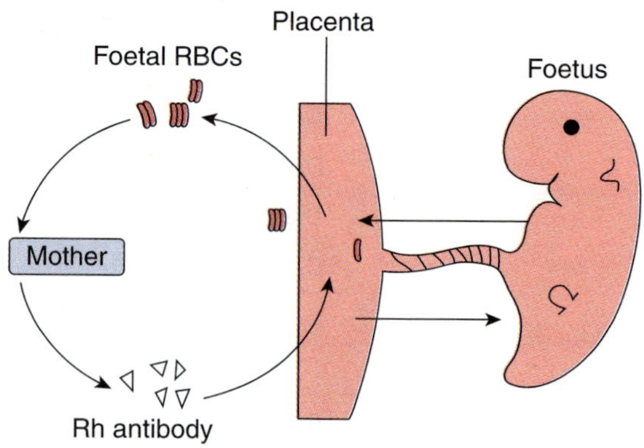

Figure 8.2 Genesis of haemolytic disease of newborn (HDN).

antibodies may get transferred across placenta to foetal circulation. They get attached to foetal red blood cells. Such anti-Rh coated cells are withdrawn from foetal circulation rendering it anaemic. To compensate this, a large number of immature red cells called "erythroblasts" are poured into foetal circulation. This offers it the name *erythroblastosis foetalis*.

Hyperbilirubinaemia may follow after birth because of the rapid destruction of red blood cells of newborn. This leads to deposition of bilirubin in the brain and, if not prevented by replacement transfusion, may cause cerebral damage. This may lead to mental retardation.

Remedial Measures

Usually, Rh sensitisation occurs at the time of delivery. At this stage (within few hours of delivery), if mother is injected with Rh immunoglobulin, it would promptly destroy Rh-positive foetal cells, before they evoke an antibody response in the mother.

OTHER BLOOD GROUP SYSTEMS

The MN blood group system was discovered by Landsteiner and Levine in 1927. On injecting human blood in rabbits, it was found that rabbit serum could then distinguish human red cells bearing M and N antigens. About two decades later, Ss subdivision of MN group was identified. MN system claims an insignificant position in transfusion or foetomaternal incompatibility.

Kell Blood Group

The original description of the Kell system identifies two phenotypes—K-positive and K-negative. It is sometimes responsible for haemolytic disease in the newborn. Kell antigen precursor is determined by gene Xk with locus on the short arm of X chromosome.

Duffy Blood Group

It was discovered in 1950. There are three alleles: Fy^a, Fy^b and Fy^0 at Duffy locus on chromosome 1. The Fy (a-b-) phenotype is resistant to malarial parasite, *Plasmodium vivax*. This gives an advantage to FyFy genotypic individuals.

Xg Blood Group

It was identified in 1962 by Mann. It is used as an X-linked marker. It is determined by Xg locus on the short arm of X chromosome. It is among the few loci on X chromosome that are not involved in X inactivation.

Lutheran Blood Group

It forms the first example of autosomal linkage and crossing over in human beings. In 1951, Mohr shown that Lutheran blood group and secretor loci were linked. The loci are located on chromosome 19. Among the best known Lutheran genes are Lu^a and Lu^b. The Lu^a gene is relatively less frequent as compared to Lu^b gene.

OTHER POLYMORPHISMS IN BLOOD

AN 75.4

Polymorphism is the occurrence in a population of two or more genetically determined forms, each with such frequencies that the rarest of them cannot be maintained by mutation only.

They include red cell enzymes. For example, glucose-6-phosphate-dehydrogenase forms one of the very good examples of polymorphism. It serves as an X-linked marker trait. The acid phosphatase (ACP) of red cells forms another polymorphism in blood having six phenotypes.

Haptoglobin

AN 75.4

It belongs to a globin class of proteins that binds with haemoglobin (Hb). It serves a job of conserving iron from red cells under destruction. Each molecule possesses two alpha (α) and two beta (β) chains. There are three variants in the α chain polymorphism. They are $H_p{}^{1F}$-fast and $H_p{}^{1s}$-slow, the third being H_p2.

Polymorphisms are also known to occur in about over 100 plasma proteins and also in DNA sequences detectable with restriction fragment length polymorphisms (RFLPs).

Summary

In 1900, Landsteiner discovered ABO blood group system. The red cell antigen forms genetic markers and is used in population studies and linkage analysis.

Clinical Applications: Include (*i*) blood transfusion, (*ii*) tissue transplantation, (*iii*) haemolytic disease of newborn (HDN)

ABO Blood Group System

There are two antigens on red blood cells, A and B. Thus, there are four phenotypes A, B, AB and O. With presence of an antigen, one expects antibodies in sera; an individual with "A" blood group shall have antibody anti-B in his serum. A person of "O" group possesses both anti-A and anti-B antibodies in his serum. A, B and O genes are at the ABO locus on the long arm of chromosome 9.

Isoagglutination: Antigen–antibody reaction is called isoagglutination. It helps to determine blood group of an individual. Subtypes of A and B have been detected as A_1, A_2 and A_{1B} and A_{2B}.

Summary—cont'd

Disease Association: Patients of gastric cancer show an excess of A group individuals. Another association exists between "O" group individuals and duodenal ulcer.

Bombay Phenotype (Oh phenotype): Identified in Bombay (1952); the Bombay phenotype shows lack of antigens A, B and H on the red blood cells as well as secretions (e.g. saliva). The serum contains anti-A, anti-B and anti-H antibodies. Individuals are homozygous for gene h. Genotype is hh.

Secretor Status: The ability to secrete A, B and H antigen is governed by secretor gene "Se" on chromosome 19q. The secretors have genotype "SeSe", while non-secretors have genotype "sese".

Rhesus Blood Group System

Named after Rhesus monkey used in experiments. Rh system is important for its role in HDN. Rh-positive persons are heterozygous or homozygous for gene coding antigen D and have antigen D on RBCs. Rh locus is on chromosome 1. There are *five Rh antigens* D, C, E, c and e with eight alleles.

Rh-Null Blood Group: These individuals do not have Rh antigens on their RBCs and present with haemolytic anaemias because Rh antigens form an integral part of red cell membrane.

Clinical Importance: Rh-negative pregnant women should be given Rh-immunoglobulin injections during pregnancy and just after parturition.

Haemolytic Disease of Newborn (HDN): In this, foetal RBCs die due to antibodies formed by mother against foetal Rh antigen. It continues for about 3 months—time taken to clear maternal antibodies from newborn circulation. The Rh-positive RBCs of the foetus pass across placental breaks and evoke production of antibodies in mother. These antibodies pass across placental barrier and destroy foetal RBCs leading HDN.

Other Blood Group Systems

These are MN blood group, Kell blood group, Duffy blood group, Xg blood group, Lutheran blood group, etc.

Other Polymorphism in Blood

Haptoglobin: It is globin class of protein that binds with haemoglobin to conserve iron. Likewise polymorphisms are known in over 100 plasma proteins.

QUESTION YOURSELF*

1. Who discovered ABO blood group system?
2. Why red cell antigens are used as genetic markers?
3. What are the phenotypes in ABO blood group system?
4. Where is the ABO locus situated?
5. What is the status of A, B and O genes in relation to each other?

6. Which blood group is presumed to be universal donor and recipient?
7. What is Bombay phenotype?
8. Where is Rh locus situated?
9. Why Rh-null blood group persons present with haemolytic anaemia?
10. What is haemolytic disease of newborn (HDN)?

*See pages 263–264 for Answers.

9 Immunogenetics

LEARNING OBJECTIVES

At the end of this chapter the students should be able to understand

- Basis of cellular and humoral immunity
- Basis of HLA system
- Basis of immune system disorders

KEY WORDS

Immunoglobulins, Linkage equilibrium, Transplantation, Complement SCID (severe combined immunodeficiency)

It deals with the genetic basis of the immunologic phenomenon in an organism. For survival and to combat effectively the antigenic challenge faced by an individual, he/she is provided with an immune system. This allows him to recognise self and non-self. The latter includes any foreign material/particle/bacterium or virus. This serves as an antigen and is taken care of by the immune system. The immune system comprises two components:

1. **Cellular immunity:** Conferred by T-cells, these are thymus-dependent cells.
2. **Humoral immunity:** Conferred by the formation of antibody produced by B-cells; these are bursa-dependent cells. Bursa of Fabricius is an organ in birds responsible for the formation of these cells.

Both T- and B-cells are derived from the same source. The T-cells are recognised according to their functional assignment as helper cell, suppressor cell and killer cell. They are responsible for delayed hypersensitivity such as rejection of grafts or delayed skin reactions. Now let us learn some of the important terms in this respect:

- *Antigen*: The substance that evokes an immune response.
- *Antibody*: It is immunoglobulin formed in response to antigenic stimulus and reacts selectively with the same antigen.
- *Immune reaction*: It is an interaction between antigen and antibody.

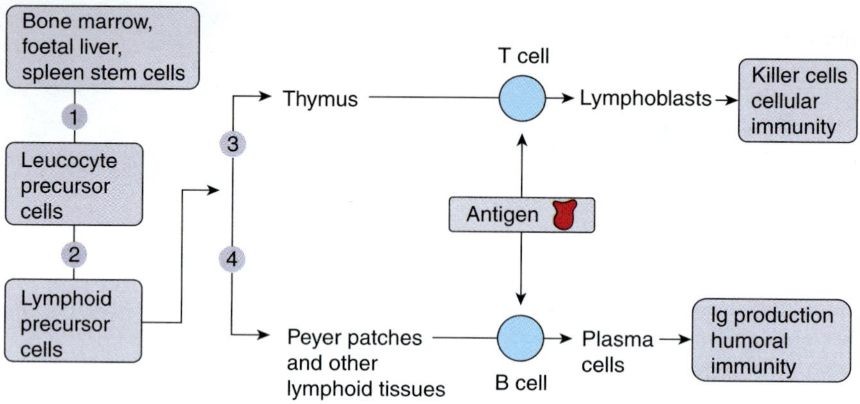

Figure 9.1 Immune system. Possible sites of blocks in immunodeficiency diseases. (1) Reticular dysgenesis; (2) Severe combined immunodeficiency (SCID); (3) DiGeorge syndrome; (4) Bruton type of agammaglobulinaemia.

The immune response of an individual with the first exposure to an antigen is called *primary response*; it takes relatively longer time, may be few days. Subsequent exposure to the same antigen evokes a rapid and more pronounced response, so-called *secondary response*. This is possible through "memory cells" lymphocytes that are primed to act vigorously with re-exposure to the same antigen. Fig. 9.1 shows genesis of the T- and B-cells and their role in immune system disorders. Of prime importance now is a question about how a body responds to "n" number of antigenic substances? This will be better elucidated as we study the immunoglobulins.

IMMUNOGLOBULINS

These are serum proteins making a fraction of about 20% of total plasma proteins. They are produced by differentiated form of B lymphocytes called plasma cells. According to available estimates, a mouse can synthesise about 10^7 to 10^8 types of antibodies. In human beings, the figure goes still higher to 10^9.

Structure

A molecule of immunoglobulin consists of four polypeptide chains. Two identical light (L) chains and two identical heavy (H) chains. These are held together by disulphide bonds. The whole unit (molecule) presents a Y-shaped structure. The light chains are kappa (κ) and lambda (λ), and the heavy chain for a particular type of Ig is single. Table 9.1 provides information about the immunoglobulins.

The heavy chains for various immunoglobulins are IgG – γ; IgA – α; IgM – μ; IgD – δ; and for IgE – ϵ. An immunoglobulin molecule can be cleaved by papain, a proteolytic enzyme, into three fragments that can be separated by chromatography. Two fragments are identical and have antigen

Table 9.1 Human Immunoglobulins

Ig Class	Heavy Chain	Light Chain	Molecular Weight	Fraction in %	Placental Transfer
IgG	γ	k or λ	150,000	80	Yes
IgA	α	k or λ	160,000	13	Possible
IgM	μ	k or λ	900,000	6	–
IgD	δ	k or λ	180,000	1	–
IgE	ϵ	k or λ	190,000	Traces	–

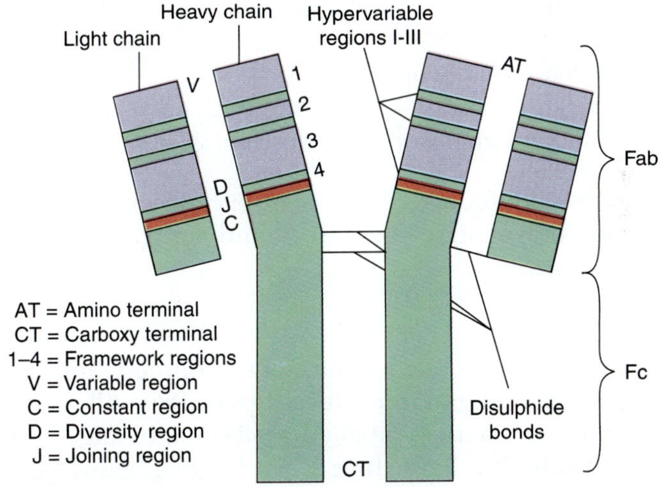

Figure 9.2 Schematic representation of antibody molecule showing generation by somatic recombination of V, D, J and cDNA segments.

Animation 4: Immunoglobulin structure. (*Available online at* www.medenact.com)

binding sites. They are called fragment–antigen–binding (FAB). The third fragment activates complement and is called FC (Fig. 9.2).

Genetically determined variants of the five classes of immunoglobulins recognised so far are:
Gm system: Associated with heavy chain in IgG
Am system: Associated with heavy chain in IgA
Km and Inv system: Associated with kappa light chain
Oz system: Associated with lambda light chain

Diversity of Immunoglobulins

The prime factor accounting for diversity in antibodies appears to be multiple combinations of heavy and light chains. A study of Bence Jones protein in patients with multiple myeloma revealed that they have two regions:

1. Variable region (V)
2. Constant region (C)

Region V further presented four regions, which did not vary much from one antibody to another. These were called *framework regions.* In between these, there were three regions that showed remarkable variations. These were called *hypervariable regions.* The DNA studies of antibody-producing cells have revealed that DNA segments coding for V and C regions of light chain are separated by J region or joining region.

The DNA sequencing of heavy chain genes revealed that they are coded by four different DNA segments, one for each of its V, D, J and C regions. The diversity region intervenes between V and J regions of heavy chain.

Chromosomal association

Gene families coding for polypeptide chains of immunoglobulins are associated with the following autosomes:

1. H : Heavy chain on chromosome 14
2. K : Kappa light chain on chromosome 2
3. λ : Lambda light chain on chromosome 22

Synthesis of immunoglobulins stands as an exception to the generalisation of "one gene–one polypeptide" because V and C regions of each chain are coded by different genes.

Class switching

On exposure to antigen, B cell produces IgM antibody initially. On subsequent exposure to this antigen it produces IgA or IgG, still retaining specificity to the same antigen. This is labelled as class switching. Further analysis reveals that the antibodies have their V (variable) regions same and differ only in C regions.

HYBRIDOMA AND MONOCLONAL ANTIBODIES

Hybridoma can be obtained by fusing two cell components:

1. *Mutant mouse myeloma cell:* These are not capable of producing antibody.
2. *Normal spleen cells from mice immunised with specific antigen:* When put to appropriate conditions, they fuse to form hybrids. Hybridoma secretes a single antibody of a single Ig type, specific for the antigen that was originally used to immunise mice. This is monoclonal antibody.

Applications

1. Hybridoma and monoclonal antibodies have helped a great deal in histocompatibility studies.
2. Gene mapping for k, λ light chains and γ, μ and α heavy chains was done making use of hybridomas.
3. Production of human immunoglobulins by fusion of mouse myeloma cells with human lymphoblastoid cells (Fig. 9.3).

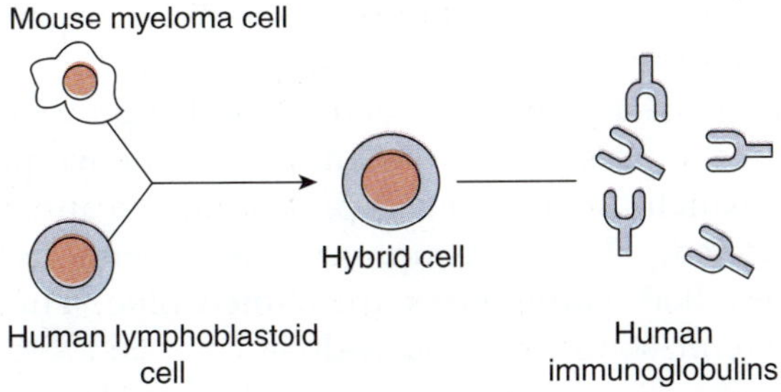

Figure 9.3 Hybridoma produced by using human lymphoblastoid cell line.

TRANSPLANTATION

In the initial phase, transplants/grafted tissue used to survive for only few days. Grafts between monozygotic twins, and to a certain extent in dizygotic twins, were usually accepted. In simple words, the tissue of genetically different sources shall encounter a reaction in recipients. This is because of antigenic disparity between the donor and the recipient. The genetic basis of rejection of graft/transplant has now been understood. Attempts have also been made to mitigate the reactions by the host towards graft helping survival of transplanted tissues.

Types of Grafts

1. **Autograft:** It is graft of the host's own tissue.
2. **Isograft:** It is graft from a genetically identical person, e.g. monozygotic twin.
3. **Allograft:** It is graft from a genetically non-identical person, and hence, graft will be rejected unless host reaction to the graft is mitigated.
4. **Xenograft:** Graft between different species. These are rejected soon.

Autograft and isograft tend to be accepted easily and allograft may be accepted with due caution. Gallico et al. (1984) reported that skin biopsy tissue could be expanded under favourable environment increasing its surface area. The same could be used in severe cases of burns.

Histocompatibility

Success of transplant depends on the so-called histocompatibility—antigenic similarity—between donor and recipient. This can be assessed by the following tests:

1. **Mixed lymphocyte culture test:** In this test, lymphocytes from both the donor and recipient are mixed in vitro. Here, lymphocyte serves in two ways:
   (a) As a responding cell (host cell) by DNA synthesis and enlargement.
   (b) As a mitogenic agent (donor cells) by stimulating mitosis.

The test offers idea about the degree of antigenic disparity between the donor and the host.

2. **Lymphocyte cytotoxicity test:** In this test, the lymphocytes are incubated with antisera in the presence of trypan blue. If the lymphocyte possesses an antigen for which the antibody is present in the antisera, the lymphocyte shall be killed. The live lymphocyte is impervious to trypan blue. Therefore, the killed lymphocytes are stained blue. This test offers idea about the antigenic status of an individual.

HUMAN LEUCOCYTE ANTIGEN (HLA) SYSTEM

It consists of four closely linked loci associated with the short arm of chromosome 6. These are arranged as D (and D related/DR), B, C and A (Fig. 9.4). In the close proximity of this HLA region is the immune response locus called *Ir locus.* The alleles associated with the various loci of the HLA region are expressed as under:

D locus: 22 alleles
B locus: 42 alleles
C locus: 8 alleles
A locus: 20 alleles

Considering the number of alleles, millions of phenotypes are possible. So the HLA phenotype of two unrelated individuals is less likely to be identical. The four loci being closely linked are inherited "en-bloc". The set of these HLA genes on each chromosome 6 constitutes *haplotype.*

Considering parental haplotypes:

Father
I and J

Mother
K and L

The possible combinations would be IK, JK, IL and JL. This means that an individual will have a 1 in 4 chance of having similar HLA antigen with a sib. Therefore, brother or sister is usually selected as a donor. Advent of immunosuppressive therapy, however, has made it possible to take care of the antigenic disparity.

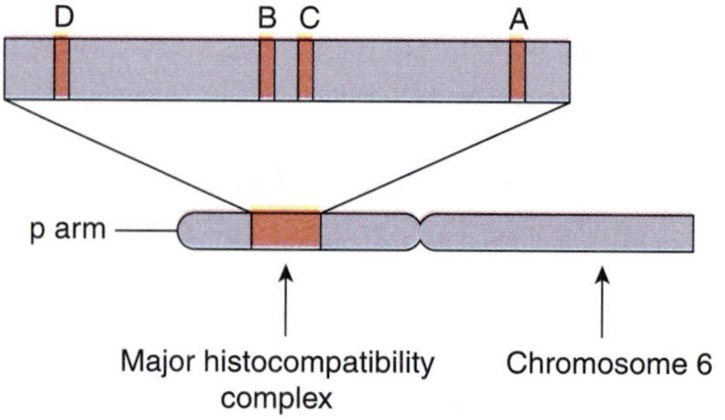

Figure 9.4 HLA complex on short arm of chromosome 6. It forms the major histocompatibility complex in man.

Linkage Disequilibrium

In the normal course of events, alleles at linked loci should follow an equilibrium. In other words, the proportions of combinations of alleles should be the product of their population frequencies. Considering the frequencies of HLA-A1 and HLA-B8, the combination of their frequencies will be:

$$\text{HLA–A1 frequency } 0.17 \times \text{HLA–B8 frequency } 0.11$$

$$\text{Al} \times \text{B8} = 0.17 \times 0.11 = 0.019$$

The product appears to be much below the observed frequency, i.e. 0.088. Thus one can say that these HLA alleles are an example of linkage disequilibrium.

Although some HLA types seem to have an association with certain disease states (Table 9.2), this does not mean that an individual possessing these HLA types shall develop the disease. It only puts the person at a relatively higher risk of developing that disease than the population in general.

Ankylosing spondylitis

Ankylosis means fusion. In this disease, mainly sacroiliac joints are involved. There is inflammation and ossification of ligaments causing fusion. In 90% of the patients with ankylosing spondylitis, HLA-B27 is present. In general population, the frequency of B27 antigen is merely 8%. In short, the relative risk (for persons with B27 antigen) is almost 100-fold in ankylosing spondylitis.

H-Y Antigen

It is Y-linked histocompatibility antigen. It is important for testicular differentiation and function. However, its expression does not depend upon presence or absence of the testicular tissue. The gene for H-Y antigen in humans

Table 9.2 HLA in Some Disease States

Disease	Antigen
Ankylosing spondylitis	B27
Chronic hepatitis	B8
Diabetes mellitus (insulin independent)	DR3, B8
Hodgkin disease	B18
Myasthenia gravis	B8, A2
Reiter syndrome	B27
Rheumatoid arthritis	DR4
Thyrotoxicosis	DR3

has been mapped to chromosome 6. It seems that Y chromosome bears a regulatory gene that governs H-Y antigen gene on chromosome 6. In experimental animals, H-Y antigen is supposed to play an important role in transplantation, but it has a little significance in humans.

Complement

These are a group of serum proteins required to inactivate foreign material after the formation of an antigen–antibody complex.

Functions

1. Direct lysis of target cells.
2. Opsonisation: Facilitating destruction of bacteria.
3. Activation of other components of the immune system.

Complement system itself has several components. Of these, four have complement loci—Bf, C2, C4A and C4B. They are closely linked to one another and are a part of major histocompatibility complex (MHC) on chromosome 6 along its short arm. These form the so-called *complotype.* One of the genetic disorders with deficiency of inhibitor of the activated C1 component is hereditary angioneurotic oedema. It is an autosomal dominant disorder.

IMMUNE SYSTEM DISORDERS

The disorders involve disruption of the immune mechanism. This may be in one of the following ways:

1. Disorders of phagocytic function (which destroy foreign particles).
2. Disorders involving T-cell and B-cell functions.
3. Lack of complement.

Chronic Granulomatous Disease

This involves phagocytosis. The phagocytes ingest the foreign particle, e.g. bacteria; but further bactericidal activity is deficient. It is an X-linked disorder. Mutation of Kell blood group locus (Xk) on X chromosome is associated with phagocytic malfunction in some cases of chronic granulomatous disease.

Reticular Dysgenesis

It is a rare autosomal recessive form of immunodeficiency. In this condition, both cellular and humoral immunity is affected. Patients also have deficiency of granulocytes. Unless treated with bone marrow transplant, these babies usually die in the first year of life.

Severe Combined Immunodeficiency or Swiss-Type of Agammaglobulinaemia

This is a severe immune system disorder that may be X-linked or autosomal recessive. When it is autosomal recessive, there is lack of adenosine-deaminase (ADA). There may be deficiency of other enzymes also, like nucleoside phosphorylase, causing an AR type of severe immunodeficiency disease. There is an abnormal function involving T- and B-cells with lowered immune response and increased chances of infection. Overwhelming infection is usually the cause of death. The management is in the form of transplantation of foetal thymus and marrow cells.

DiGeorge Syndrome

It is characterised by the absence of thymus as well as parathyroid glands. In it, there is lowered cell-mediated response due to lack of T-cell function. However, plasma cells appear to operate giving Ig synthesis.

Bruton-Type Agammaglobulinaemia

It is manifested as a B-cell deficiency with onset in infancy. It is transmitted as an X-linked recessive disorder. Thymus is normal with normal number of T-cells. Immunoglobulins are absent, although cell-mediated immunity is normal. The patient suffers from bacterial infection. It can be helped by injecting immunoglobulins and antibiotics.

Summary

Immunogenetics: It deals with genetic basis of immunological phenomenon. The immune system has two components:

- ***Cellular immunity:*** Conferred by T-cells (thymus-dependent cells). T-cells are recognised according to their function as helper, suppressor and killer cells.
- ***Humoral immunity:*** Conferred by B-cells (bursa-dependent cells)

Terms to Learn and Remember

Antigen: A substance that evokes an immune response.

Antibody: It is immunoglobulin formed in response to antigen.

Immune reaction: An interaction between antigen and antibody.

Primary response: Immune response with the first exposure to antigen.

Secondary response: Subsequent exposure to the same antigen evokes a rapid and more pronounced so-called secondary response. This is due to "memory cells".

Immunoglobulins (Ig): These are serum proteins making about 20% of plasma proteins. They are produced by B lymphocytes (plasma cells). In humans, 10^9 types of antibodies can be formed.

Continued

Summary—cont'd

Structure: The immunoglobulin molecule has four polypeptide chains:

a) ***Two light chains:*** kappa (k) and lambda (λ)

b) ***Two heavy chains:*** heavy chain for a particular type of Ig is single type, for IgG – γ; IgA – α; IgM – μ; IgD – δ and IgE – ϵ.

Immunoglobulin molecule can be cleaved by papain into three fragments. Two fragments are identical and have antigen binding sites (FAB). Third fragment activates complement (FC).

Diversity of Immunoglobulins: It is due to multiple combinations of heavy and light chains. They have two regions: (i) variable region (V), presenting four regions, called framework regions; between them there are three hypervariable regions; (ii) constant region (C).

Chromosomal Association: Genes for:

i) H: heavy chain – chromosome 14

ii) k: Kappa light chain – chromosome 2

iii) λ: Lambda chain – chromosome 22

Class Switching: On exposure to antigen, B cell produces IgM antibody initially. On subsequent exposure to this antigen, it produces IgA or IgG, still retaining specificity to the same antigen. This is called class switching.

Hybridoma and Monoclonal Antibodies: Mutant mouse myeloma cells cannot produce antibody. Normal spleen cells from mice immunised with specific antigens fuse to form hybrids when put to appropriate conditions. This hybridoma secretes single Ig specific to that antigen (used to immunise mice). This is called *monoclonal antibody*.

Applications: (*i*) In histocompatibility/studies; (*ii*) gene mapping; (*iii*) production of human immunoglobulin

Transplantation/Grafting

Types of grafts:

- ***Autograft:*** Graft of host's own tissue.
- ***Isograft:*** Graft from genetically identical person, e.g. monozygotic twin.
- ***Allograft:*** Graft from genetically non-identical person—rejected.
- ***Xenograft:*** Graft from different species.

Histocompatibility: Antigenic similarity between the donor and recipient governs the success of transplant/graft, i.e. histocompatibility between them. It is assessed by (i) mixed lymphocyte culture test and (ii) lymphocyte cytotoxicity test.

Human Leucocyte Antigen (HLA) System

It consists of four closely linked loci with following number of alleles on "p" arm of chromosome 6. They are: D locus – 22 alleles; B locus – 42 alleles; C locus – 8 alleles and A locus – 20 alleles. With these alleles, millions of phenotypes are possible; hence HLA phenotype of any unrelated persons is less likely to be identical.

Linkage Disequilibrium: Alleles at linked loci should follow an equilibrium, i.e. the proportion of combinations of alleles should be the product of their population frequencies.

HLA – A_1 frequency 0.17 and HLA B8 frequency 0.11.

A1 × B8 = 0.17 × 0.11 = 0.019. This is much less than the actual frequency 0.088. In other words, HLA alleles are example of linkage disequilibrium. Some HLA and disease associations are:

Ankylosing spondylitis – B27;
Myasthenia gravis – B8, A2;
Diabetes mellitus – DR3, B8
Rheumatoid arthritis – DR4.

Summary—cont'd

H-Y antigen: It is Y-linked histocompatibility antigen, important for testicular differentiation. Gene for it is located on chromosome 6.

Complement: It is a group of serum proteins required to inactivate foreign material after formation of antigen–antibody complex. *Functions* are: (*i*) lysis of target cells; (*ii*) opsonisation; (*iii*) activation of other components of immune system.

Immune System Disorders

i) ***Chronic granulomatous disease:*** Phagocytic malformation and an X-linked trait.
ii) ***Reticular dysgenesis:*** An autosomal recessive form of immunodeficiency.
iii) ***Severe combined immunodeficiency (SCID)/Swiss-type of agammaglobulinaemia:*** An AR/XR trait-associated abnormal T-cell and B-cell function. Lack of adenosine diaminase (ADA) or other enzyme, like nucleoside phosphorylase, may be the cause.
iv) ***DiGeorge syndrome:*** Absence of thymus and parathyroid is associated with immunodeficiency and disturbances of calcium metabolism.
v) ***Bruton-type agammaglobulinaemia:*** B-cell deficiency, transmitted as X-linked recessive disorder. Immunoglobulins are absent. Thymus and T-cells are normal.

QUESTION YOURSELF*

1. What is immunogenetics?
2. What is humoral immunity/response?
3. What is primary immune response?
4. What are immunoglobulins?
5. What is the structure of an immunoglobulin molecule?
6. Why synthesis of immunoglobulin stands as an exception to the generalisation "one gene–one polypeptide"?
7. What is class switching?
8. What are different types of grafts?
9. Which grafts are easily accepted?
10. What is meant by histocompatibility?
11. What are the tests to assess the histocompatibility?
12. What is HLA system?
13. Where is H-Y antigen located?
14. What is complement?
15. What are the types of immune responses?

16. What is an antigen?

17. What is secondary response?

18. What are memory cells?

19. What is the cause of diversity in immunoglobulins?

20. Match the heavy chain with the respective immunoglobulins.

1. IgA	a. γ (Gamma)
2. IgD	b. α (Alpha)
3. IgE	c. δ (Delta)
4. IgG	d. μ (Mu)
5. IgM	e. ϵ (Epsilon)

21. What are the two types of light chains?

22. What is the chromosomal association of the genes coding for the heavy and the light polypeptide chains?

23. What is hybridoma?

24. What is SCID?

---

*See pages 264–265 for Answers.

Cancer Genetics 10

LEARNING OBJECTIVES

At the end of this chapter the students should be able to understand

- Basis of role of Genetics in cancer
- Role of oncogenes
- Mutations and chromosomal aberrations leading to cancer

KEY WORDS

Oncogenes, Retrovirus, Tumour suppressor genes, Philadelphia chromosomes

Association between cancer and human chromosomes was first thought in 1902 by Professor Theodor Boveri at the University of Würzburg in Germany. He stated that a change or changes in one or more chromosomes might be the starting point for the development of cancer. It took more than half a century to substantiate his theory. In 1959, Peter C. Nowell and David A. Hungerford opened a new page in cancer research with the discovery of Philadelphia chromosome. Over the years, cancer has remained a mystery to the scientific world. With numerous carcinogens in the environment, some individuals are more prone to develop cancer while others are not. This leads us to think that human genome might play a significant role in the genesis of this mysterious disease. Moreover, the chromosome study of cancer cells often shows marked variation in number, as well as structure of chromosomes. Analysis of variety of tumour cells reveals that certain chromosomes are involved in certain cancers (Table 10.1). It is thought that they carry genes that bring about a malignant transformation.

Viral Association: There are a number of DNA tumour viruses associated with neoplastic conditions in humans, while numerous RNA viruses/retroviruses cause tumour in animals. They are shown in Table 10.2.

RETROVIRUS

In the early part of 20th century, it was recognised that certain types of cancer in chickens could be transmitted by cell-free passage. Later on, it was shown

Table 10.1 Chromosomal Localisation of Proto-oncogenes in Human Genome

Chromosome Number	Site	Gene
1	$q^{12\ ter}$	*c-src-2*
1	Cen–p^{21}	*N-ras*
1	p^{32}	*Hum* B lym^{-1}
1	NA	*c-ski*
3	p^{25}	*c-mil* (raf)
5	q^{34}	*c-fms*
6	q^{22}–q^{24}	*c-myb*
7	p^{ter}–q^{22}	*c-erbB*
8	q^{22}	*c-mos*
8	q^{24}	*c-myc*
9	q^{34}	*c-abl*
11	$P^{14.1}$	*C-ras*^{H}
11	q^{23}–q^{24}	*c-ets*
12	$P^{12.05\ -\ ter}$	*c-ras*^{K}
14	q^{21}–q^{31}	*c-fos*
15	q^{24}–q^{25}	*c-fes*
17	q^{21}–q^{22}	*c-erbA*
20	P^{34}–p^{36}	*c-src-1*
22	$q^{11\ -\ ter}$	*c-sis*

Table 10.2 Human DNA Viruses Causing Cancer

Family	Type	Tumour
Herpes	Epstein–Barr (EBV)	Burkitt lymphoma, nasopharyngeal carcinoma
Papova	Human papilloma virus (HPV)	Plantar and genital warts, urogenital cancer, skin cancer
Hepadna	Hepatitis B virus (HBV)	Hepatocellular carcinoma

(*Source: Emery's Elements of Medical Genetics,* Eleventh Edition, 2001, Table 13.1, Page 191, Churchill Livingstone.)

that this was possible due to a virus. A wide variety of these viruses were identified and their association with specific types of cancer was noted. These viruses are called retroviruses. They are peculiar in that they have RNA as their genetic material instead of DNA. When this virus infects the host cell, it makes DNA copy. It is called *reverse transcription.* The DNA copy is incorporated in the host genome. The specific DNA sequences formed by these viruses in the host cell are responsible for the neoplastic change.

Retroviral genome consists of RNA molecule with about 8000 to 10,000 nucleotides. Within host cell, it forms DNA that gets incorporated in host DNA. The integrated DNA thus formed by blending the host DNA and DNA copy formed by reverse transcription is called provirus. The structure of provirus shows long terminal repeats (LTRs) copied from viral genomic RNA at both ends. Until recently, naturally occurring retroviruses were thought to have only three genes—"*gag*", encoding the structural proteins for core antigens; "*pol*", coding for reverse transcriptase enzyme; and "*env*", the gene for the glycoprotein envelope proteins (Fig. 10.1). A study of Rous sarcoma virus has identified fourth gene that results in transformation of cells in culture (Fig. 10.2); this viral gene is known as *oncogene.*

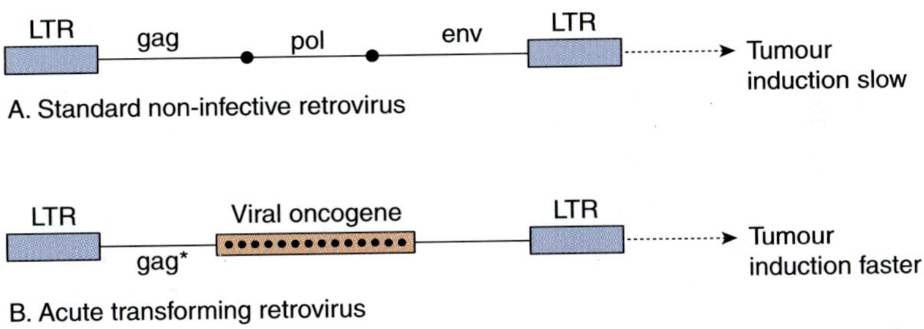

Figure 10.1 Structure of retroviruses and their effect on tumour induction (gag* indicates that a part of gag sequence is deleted). *Abbreviation*: LTR, long terminal repeats.

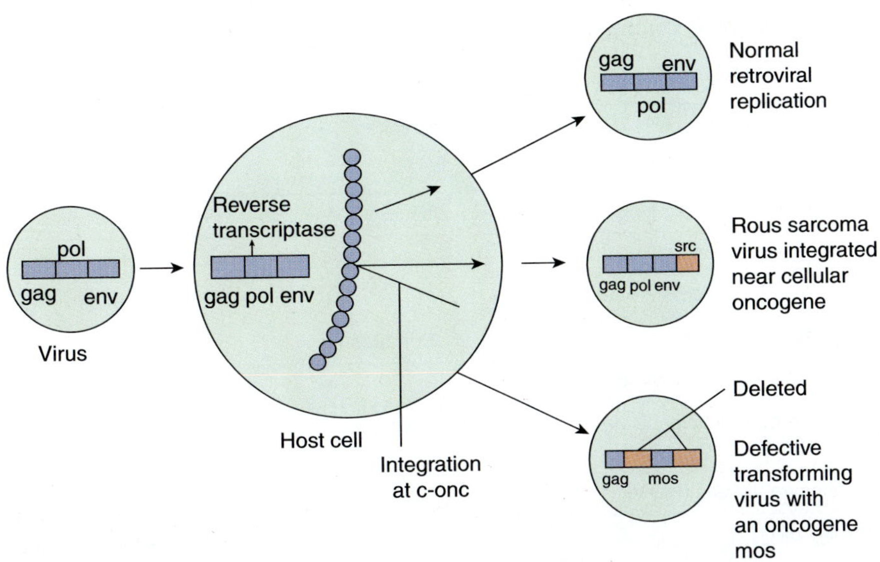

Figure 10.2 Diagram showing transforming ability of the retrovirus and the replication. *(Source: Emery's Elements of Medical Genetics,* Eleventh Edition, 2001, Fig. 13.1, Page 191, Churchill Livingstone.)

ONCOGENES

Oncogenes are recognised by three letter abbreviations, which reflect over their origin or the type of tumour with which they are associated (Tables 10.3 and 10.4). Normal mammalian cells contain DNA sequences homologous to the viral oncogenes; these have been called proto-oncogenes or cellular oncogenes. In strict sense, the term proto-oncogene refers to normal gene, while the term cellular oncogene or *C-onc* refers to a gene with oncogenic potentials like viral oncogene or *V-onc*.

V-onc Formation: The retroviral oncogenes are formed because of errors in the replication of the retrovirus genome following their integration at random sites in the host DNA (Fig. 10.2). The resultant being a viral oncogene. For example, the viral oncogene *sis* is almost identical to the gene for one chain of a growth factor called platelet-derived growth factor (PDGF).

From Proto-oncogenes to Cellular Oncogenes

A couple of models have been postulated to explain conversion of proto-oncogenes to cellular oncogenes. In the first quantitative one, tumour formation is induced by increase in the amount of proto-oncogene product. In the second, there is conversion of proto-oncogenes into a transforming gene (c-*onc*) with nucleotide sequence being altered, i.e. a qualitative.

Table 10.3 The Oncogenic Retroviruses and the Associated Tumours

Virus	Host	Tumour
Rous sarcoma virus	Chicken	Sarcoma
Avian leukosis virus		Avian leukaemia
Murine sarcoma virus	Mice	Sarcoma
Murine leukaemia virus		Leukaemia
Mouse mammary tumour virus		Breast cancer
Simian sarcoma virus	Primates	Sarcoma
Gibbon ape leukaemia virus		Leukaemia
Human immunodeficiency virus type 1 (HIV-1)	Humans	Kaposi sarcoma
Lymphocyte viruses (HLTV)		
Human cell		T-cell leukaemia

(*Adapter from: Emery's Elements of Medical Genetics*, Eleventh Edition, 2001, Table 13.3, Page 192, Churchill Livingstone.)

Table 10.4 Transforming Viruses, Species Affected, an Oncogene Responsible and the Tumour Induced

Virus	Species	Oncogene	Tumour
Avian erythroblastosis	Chicken	*erb*-B	Erythroleukaemia
Avian myeloblastosis	Chicken	*myb*	Myeloblastic leukaemia
Avian myelocytomatosis	Chicken	*myc*	Myelocytoma, sarcoma
Abelson leukaemia	Mouse	*abl*	Pre-B cell leukaemia
FBJ murine osteosarcoma	Mouse	*fos*	Osteosarcoma
Moloney murine sarcoma	Mouse	*mos*	Sarcoma
Harvey murine sarcoma	Rat	*Ha-ras*	Sarcoma
Kirsten	Rat	*ki-ras*	Sarcoma
Rous sarcoma	Chicken	*src*	Sarcoma
Simian sarcoma	Monkey	*sis*	Sarcoma

Quantitative model of oncogenic function

Two mechanisms can give rise to an inappropriate amount of proto-oncogene products, gene amplification and insertional mutagenesis.

Gene Amplification

The mechanism of "survival" is responsible for it. For example, leukaemic cells, when exposed to the chemotherapeutic agent methotrexate, acquire resistance by making multiple copies of the gene for dihydrofolate reductase. It is target enzyme for methotrexate. The gene amplification increases number of copies of the oncogene several times, thus corresponding oncoprotein product also increases. In mammals, an amplified DNA sequence in tumour cells can be evidenced as small extra chromosomes called ***double-minute chromosomes/homogeneously staining regions (HSRs)*** of the chromosomes. HSRs are found in many tumour cells. N-*myc* is amplified in about 30% of neuroblastomas. In advanced cases, the figure tunes to 50%. In human small cell carcinoma of the lung, amplification of *c-myc*, N-*myc* and L-*myc* is found. Amplification of c-*neu* or *erb*-B_2 is encountered in 20% of breast cancers.

Insertional Mutagenesis

The retrovirus can transform a cell without V-*onc* expression. This was first noted in a study of bursal lymphomas caused by transformation of B-lymphocytes with avian leukosis virus. When such a retrovirus integrates into the host genome close to a proto-oncogene, the viral DNA long terminal repeat sequences have a potential to induce uncontrolled expression of the cellular gene. The resulting entity is called *insertional mutagenesis.* It has been suggested that c-*myc* activation in this manner probably leads to the development of Burkitt lymphoma in humans, which is associated with Epstein–Barr viral infection.

Qualitative model on oncogene function

In addition to the two methods described in quantitative model, there are two other ways also by which the proto-oncogenes can be converted into the cellular oncogenes.

Mutations in Coding Sequences

A cell line derived from mouse fibroblasts called NIH 3T3 was transformed by DNA transfection from a human bladder carcinoma cell line. This led to the discovery of a human DNA sequence homologous to the "*ras*" gene of the Harvey murine sarcoma virus. The human "*ras*" gene family comprises of three closely linked members, H-*ras*, Ki-*ras* and N-*ras*. The ras proteins are structurally similar to their viral counterparts except near their carboxyl terminals. The oncogenic potential of the ras proto-oncogenes is because of point mutations in their nucleotide sequence.

The activating mutations in the *ras* gene have been detected in about 30% of the human cancers; however, the incidence depends upon the tumour origin. For example, *ras* oncogenes are found in 25%–30% of lung cancers, 50% of colonic cancers, 90% of the pancreatic carcinomas, but are almost non-existent in breast cancers. Such activating mutations in the "*ras*" gene are also found in some premalignant lesions suggesting their possible role in initiating the neoplasm.

Chromosomal Aberrations

The chromosomal aberrations are frequently encountered in malignant cells. They may be in the form of variation in the number or the structure of chromosomes. They are as follows:

Tumour Suppressor Genes

In late 1960s, Harris and his colleagues carried out fusion of malignant cells with normal cells in culture. They found that malignant phenotype was suppressed in the hybrid cells. This was attributed to the action of tumour suppressor genes in the normal cells. Loss of this tumour suppressor activity leads to malignancy. Often these genes are called anti-oncogenes.

Role of Tumour Suppressor Genes: Familial retinoblastoma has been considered as an autosomal dominant trait. However, demonstration of retinoblastoma gene (Rb gene) has given a concept that it is a recessive tumour suppressor gene. An absence of the gene product in the homozygous state leads to the development of this peculiar tumour. The tumour suppressor activity of Rb gene is also demonstrated in vitro in cancer cells.

Retinoblastoma (Rb) Gene: The Rb gene encodes p105, a nuclear protein that is associated with DNA and involved in the regulation of the cell cycle.

It is believed that p105 is also involved in regulating transcription of critical genes. The Rb gene product exists in two conditions—phosphorylated and unphosphorylated. In its latter state (unphosphorylated), it is inactive in growth suppression, while in its phosphorylated state it gets associated with an unknown/unidentified nuclear factor and suppresses growth. The Rb gene product interacts with several viral oncoproteins, e.g. transforming proteins of SV40 and papilloma virus (E7 protein).

p53 Gene: The p53 protein was first identified as host cell protein bound to T-antigen. On cloning murine p53, it was shown to be able to cooperate with an activated *ras* and serve as an oncogene to transform the cells in vitro, though the rodent cells expressed normal or wild type p53. An inactivation of p53 was often found in murine Friend virus induced erythroleukaemia cells. This led to the concept that p53 is a tumour suppressor gene.

The p53 mutations are found in 50% of bladder, breast, colon and lung cancers. This involves highly conserved regions in exons 5 to 10. As against this, p53 mutations in hepatocellular carcinoma involve a "hot spot" in codon 249, the base change being from G to T. This could be as a result of an interaction with aflatoxin B, a carcinogen, associated with hepatic cancer in China. Hepatitis B virus may serve as a co-factor.

The p53 protein is a multimeric complex, with its mutant monomers more stable than normal p53 protein. The mutant monomers can form complexes with the normal wild type p53 to inactivate it, acting in a *dominant negative manner.* The p53 has been recognised as "guardian of the genome", since it allows DNA damaged through normal wear and tear to be repaired by preventing progress of the cell cycle.

Li-Fraumeni Syndrome. This rare entity is inherited as an autosomal dominant trait. Persons with this condition have point mutations in highly conserved regions of p53 gene (codons 245–258) involving the germ cell line of family members. These individuals are highly susceptible to a variety of early onset malignancies, which include breast cancer, adrenal carcinoma, sarcomas etc.

Deleted in Colorectal Cancer (DCC): The DCC gene is expressed in the normal colonic mucosa and is remarkably reduced or absent in colorectal carcinoma. The loss of DCC plays a role in cell-to-cell and cell-to-basement membrane interactions, the feature obviously found in malignancy. It is associated with 18q, the loss of which is found in over 70% of colorectal cancers.

Classification and Functions of Oncogenes

Oncogenes are classified according to their cellular location and the function of the oncoproteins encoded by them in the signal transduction pathway (Table 10.5). The growth factors stimulate cells to grow by binding to one of the three types of specific growth factor receptor in a process called signal transduction. The three growth factor receptor types are shown in Table 10.6.

Table 10.5 Classification of Oncogenes

Nuclear Oncogenes	Cytoplasmic Oncogenes	Post-Receptor Tyrosinase	GTP Binding	Growth Factor Receptors	Growth Factors
fas	*mos*	*src*	*N-ras*	*erb-B*	*sis* (PDGF)
jun	*A-raf*	*abl*	*Ha-ras*	*erb*-B2	*int* (FGF-related)
erb-A	*B-raf*	*yes*	*ki-ras*	*fms*	*hst* (FGF-related)
myb		*fgr*		*Kit*	
myc		*fes*		*ros*	
N-myc		*syn*		*trk*	
L-myc				*ret*	

(*Adapted and Modified from: Emery's Elements of Medical Genetics*, Eleventh Edition, 2001, Table 13.4, Page 193, Churchill Livingstone.)

Table 10.6 Types of Growth Factor Receptors with Corresponding Growth Factors

Receptor	Growth Factor
GTP binding proteins	Thrombin, serotonin angiotensin
Protein tyrosine kinase	Insulin, PDGF, EGF
Cytoplasmic tyrosinase kinase-linked haemopoietin receptors	Prolactin, erythropoietin

Growth factors

These are substances governing transition of a cell from G0 phase till the starting of the cell cycle. Different cells need different growth factors to stimulate cell division. Two well-known and -studied growth factors are PDGF and EGF.

Platelet-derived growth factor (PDGF): It stimulates the proliferation of the connective tissue cells.

Epidermal Growth Factor (EGF): It stimulates number of cell types including epidermal cells. The best known oncogene that serves as a growth factor is *sis* oncogene. It encodes part of "b" chain of PDGF. When PDGF is added to non-tumourigenic long-term cell cultures, such as NIH 3T3, the cells are transformed. They behave like neoplastic cells. Their growth rate increases. In vivo, they form tumour; in vitro, they lose contact inhibition. A couple of oncogene products showing homology to fibroblast growth factors are *int*-2 and *hst*. They are amplified in malignant melanomas, breast and oesophageal cancers.

The **signal transduction** earlier referred to is a complex multistep pathway extending from the cell membrane via cytoplasm to nucleus. It has both positive as well as the negative feedback loops required for an accurate proliferation and differentiation of the cell.

Nuclear oncogenes

The oncogenes *fos, jun* and *erb*-A encode proteins that regulate gene expression by activating or suppressing DNA sequences close to it. The *c-myc* probably relates to alterations in control of the cell cycle. The *c-myc* and *c-myb* oncoproteins stimulate progress of the cells from G1 to S phase in the cell cycle. Their overproduction does not allow the cells to enter prolonged resting phase. This results in cellular proliferation.

Cytoplasmic oncogenes

Numerous cytoplasmic gene products form a part of signal transduction pathway. The *"raf"* gene product modulates the normal regulating cascade. It may be directly responsible for transmitting growth promoting signal to the nucleus.

Post-receptor tyrosine kinases

In normal cells, the phosphorylation of amino acid tyrosine is uncommon. This phosphorylation is done by gene products that are involved in signal transduction. Both, the *abl* oncogene and *src* oncoprotein have tyrosine kinase activity. The *src* oncoprotein is responsible for the transforming properties of the Rous sarcoma virus.

GTP binding proteins

There are intracellular proteins associated with protein kinases. The *ras* proteins are located on the inner surface of the plasmalemma.

Growth factor receptors

Many oncogene products (oncoproteins) have tyrosine kinase activity. The receptors for tyrosine kinases include c-*erb*-B. It encodes the epidermal growth factor (EGF) receptor and is related to *c-erb*-B_2 oncogene. This oncogene *c-erb*-B_2 is amplified and expressed in gastric, pancreatic and ovarian cancers.

Apoptotic oncogenes

The *bcl*-2 oncogene is an example of apoptotic oncogene. The cancer cells increase in number enormously owing to increased growth and/or division or decreased cell death. Programmed involution or cell death is called apoptosis.

Familial cancer

It is believed that about 5% of the colorectal and breast cancer arises as a result of an inherited cancer susceptibility gene. There are a number of other features that suggest an inherited cancer susceptibility syndrome in a particular family. These features include the following:

1. Several first- or second-degree relatives show a common cancer.
2. Several close relatives have related cancers, e.g. breast and ovary, or bowel and endometrial cancer.
3. Early age of onset.
4. Tumours in two different organ systems in one person.
5. Bilateral tumours in paired organs.
6. Two family members may have the same rare cancer.

Familial cancer-predisposing syndrome

Most cancers occur at a specific site, but there are some families in which the cancers occur at different sites in various members of the family or at more than one site in an individual. The incidence is more common than would be expected. Such families are referred to have familial cancer-predisposing syndrome. Most of them are with dominant inheritance. Some of them are listed in Table 10.7.

Clinical Implications of Oncogenes

1. Study of the neoplastic changes at the molecular level in different cell lines from tumours will help in early diagnosis. For example, alterations, involving c-abl RNA and c-abl coded protein in a chronic myelogenous leukaemia case, can be studied at an early stage.
2. If we know precisely the modus operandi of oncogenes at the molecular level—how they bring about transformation in cells, then one can think of an effective drug intervention.

Table 10.7 Inherited Familial Cancer Predisposing Syndromes

Syndrome	Inheritance	Chromosomal Location	Gene	Cancer
Breast and breast/ ovary families	AD	17q	BRCAI	Breast, ovary
Familial adenomatous polyposis	AD	5q	APC	Colorectal, duodenal, thyroid
Familial retinoblastoma	AD	13q	Rbl	Retinoblastoma
Li-Fraumeni syndrome	AD	17p	p53	Sarcoma, breast, brain, leukaemia, adrenal cortex
Multiple endocrine neoplasia (MEN) Type 2a	AD	10q	MEN2a	Thyroid, pheochromocytoma

(*Modified and Adapted from: Emery's Elements of Medical Genetics*, Eleventh Edition, 2001, Table 13.6, Page 199, Churchill Livingstone.)

CHROMOSOMAL ABERRATIONS IN CANCER

They can be in the form of:

1. Translocations
2. Deletions
3. Chromosome breakage syndromes

Translocations

Under this category, there are two well-recognised disorders. They are as under:

Chronic myeloid leukaemia (CML): Philadelphia chromosome

In 1960, Peter C. Nowell and David A. Hungerford discovered a small chromosome in patients with chronic myelogenous leukaemia. Later on, it was called Philadelphia chromosome. Formation of Philadelphia chromosome involves translocation between the long arm of chromosome 22 and the long arm of chromosome 9 (Fig. 10.3). This results in shorter chromosome 22q- (Ph^1) called Philadelphia chromosome. It has also been found in few cases of acute lymphoblastic leukaemia.

Detailed consideration

Experiments of chromosome mapping have shown c-abl (identified from Abelson murine leukaemia virus), a cellular oncogene on the long arm of chromosome 9, which is involved during Philadelphia translocation. Similarly on the long arm of chromosome 22, an oncogene *c-sis* has been mapped out. In situ hybridisation studies show that these two oncogenes reside close to the chromosomal breakpoints and are included within exchanged segments. The exact function of c-abl is unknown, but it belongs to a family of genes encoding proteins called tyrosine kinases, which play a role in cell growth. The oncogene *c-abl* is translocated to the "*bcr*" gene (breakpoint cluster region) of

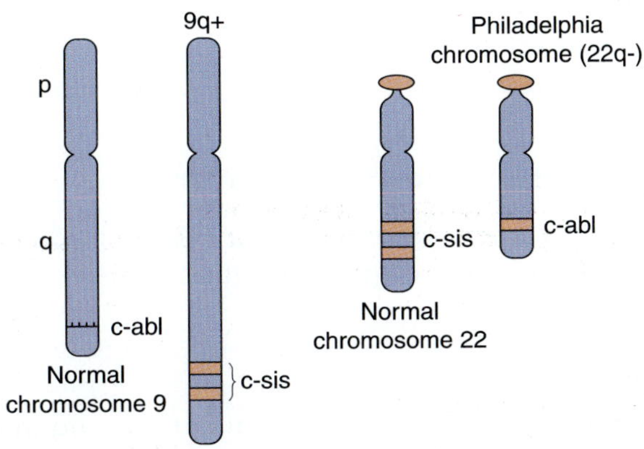

Figure 10.3 Formation of Philadelphia chromosome involves reciprocal translocation between q arms of chromosome 9 and 22.

chromosome 22. A giant RNA molecule is transcribed from the fused gene and is then spliced into the messenger RNA. This messenger RNA is then translated into fusion protein. This protein contains about 900 amino acids of *bcr* (break point cluster region) and 1100 amino acids of *c-abl*.

Clinical significance

In 1968, the bone marrow study of CML patients was carried out. The patients were divided under two heads—Ph-positive (having Philadelphia chromosome) and Ph-negative cases. In Ph-negative patients, the average survival period was 12–15 months; in Ph-positive cases, the average survival period was 36–44 months. Yet another study conducted in 1972 showed Ph chromosome in 19%–22% cells in individuals who developed chronic myelogenous leukaemia after 5½ years. In other words, in asymptomatic individuals, presence of Ph chromosome indicates a preleukaemic state.

Future prospects

Fusion protein of chronic myelogenous leukaemia has a unique structure, which means that it will be possible to raise an antibody against it. Once it is possible, then this antibody could be used for early detection of CML and also in the treatment of the disease.

Burkitt lymphoma

It is prevalent in Central Africa and commonly involves children. It is a neoplastic disease involving lower jaw. It is characterised by an osteolytic lesion. In this, there is translocation involving chromosome 8 and 14. Translocation transfers *c-myc* gene from chromosome 8q24 to 14q32. Thus, translocation involves c-*myc* gene on chromosome 8 and immunoglobulin heavy chain gene locus on chromosome 14. It is believed that translocation increases transcription of *c-myc* almost to 20 times. Alternatively, in some patients the gene product may be altered instead of being increased. In some cases of Burkitt lymphoma, the breakpoint on chromosome 8 remains the same, but the reciprocal translocation involves the short arm of chromosome 2 near locus of k light chain immunoglobulin gene or the long arm of chromosome 22 near λ light chain locus.

The exact mechanism by which translocation of *c-myc* close to immunoglobulin gene causes neoplastic transformation is still obscure.

Deletions of Chromosomes and Cancer

Some forms of cancer are associated with a loss of chromosome material, i.e. deletion. Few of the well-recognised entities have been described in the following part:

Retinoblastoma (Rb)

It is an embryonic tumour involving the retina. It is found in children and is often inherited as an autosomal dominant disorder. Rb locus has been mapped out along the long arm of chromosome 13 (designated 13q14). In few children with retinoblastoma, there is interstitial deletion involving this part of chromosome 13. In patients with retinoblastoma, a comparative study of restriction mapping of DNA is done from tumour cells and normal cells. This reveals a variety of secondary somatic events in homologous chromosome 13. These are in the form of mutation or deletion of normal allelic gene producing retinoblastoma.

Aniridia-Wilms tumour

Wilms tumour is a malignant tumour involving kidney. It occurs in children and may start even prenatally. A proportion of the suffering children may show deficient iris (aniridia) associated with kidney tumour. About 50 such cases have been reported so far. Many of these patients also show a mental deficiency and growth retardation. A cytogenetic study shows interstitial deletion involving the short arm of chromosome 11 (i.e., 11p–). The manifestations seen are because of partial monosomy of the short arm of chromosome 11. One of the oncogenes, c-ras^H (Harvey murine sarcoma), has been mapped in this portion of chromosome 11.

Chromosomal Breakage Syndromes

On cytogenetic analysis of a number of autosomal recessive disorders, we find excessive chromosomal instability, i.e. breaks or gaps are evident along chromosomes. These individuals have defective DNA repair mechanisms and are more prone to malignancy.

Fanconi anaemia

It is an autosomal recessive disorder. There are associated upper limb abnormalities. It is characterised by pancytopenia owing to bone marrow failure. The skin shows brownish pigmentation. Upper limb malformations include hypoplastic thumb and severe defects such as radial aplasia. There is often prenatal and postnatal growth retardation. The haematological picture shows increased level of foetal haemoglobin (HbF). Increased chromosomal breaks are evident on marrow cell cultures. Experimentally, it was found that fibroblast cultures from the patients show increased chromosome breakages in presence of mitomycin C. Finally, these patients are more prone to develop leukaemia and lymphomas.

Ataxia–telangiectasia

It is an autosomal recessive disorder presenting in late childhood with ataxia (of cerebellar origin) associated with telangiectasia (dilated blood vessels) often seen in bulbar conjunctiva and auricles. These individuals have defective cellular immunity. They also show low serum IgA levels and hypoplastic thymus. They are susceptible to pulmonary infections. They lack an ability to excise damaged DNA owing to radiation. They present an increased risk of developing malignancy.

Xeroderma pigmentosum

It is a skin disease often turning into skin malignancy. A characteristic feature of the disorder is photosensitive skin rash in sun-exposed areas. It often involves an age group below 20 years. Cultured cells from these patients show an increased frequency of sister chromatid exchanges (SCEs) on exposure to UV light or chemical carcinogens. There is deficiency of enzymes, associated with excision and repair of UV induced DNA damage.

Chromosomal instability or breakages are associated with increased frequency of SCEs in cell cultures. Normally, about 10 SCEs per cell are found, but the number is remarkably high in xeroderma pigmentosum patients.

Summary

In 1902, Professor Theodor Boveri for the first time thought of association of cancer and chromosomes. PC Nowell and DA Hungerford discovered Philadelphia chromosome in CML cases in 1959.

Retroviruses: They have RNA as genetic material instead of DNA. When this virus infects host cell, it makes DNA copy. It is called *reverse transcription*. The DNA copy is incorporated in the host genome. The specific DNA sequences formed by these viruses in the host cell are responsible for neoplastic change. The integrated DNA thus formed by blending of host DNA and DNA copy formed by reverse transcription is called provirus.

Oncogene: A study of Rous sarcoma virus has identified fourth gene, the *oncogene*, apart from earlier three known viral genes "*gag*", "*pol*" and "*env*". Oncogenes are identified by three letters, reflecting over their origin or type of tumour. Normal mammalian cells contain DNA sequences homologous to viral oncogenes; they are called *proto-oncogenes*.

Cellular oncogenes (C-onc): These genes have oncogenic potentials like viral oncogenes (*V-onc*). The viral oncogenes are formed because of errors in replication of viral genome. Two models have been postulated to explain conversion of proto-oncogenes (normal) to cellular oncogenes (with oncogenic potential):

1. *Quantitative model* of oncogenic function: Two mechanisms can do so—gene amplification and insertional mutagenesis.
2. *Qualitative model* of oncogenic function: The proto-oncogenes can be converted into cellular oncogenes by one of the following two ways:
   i) *Mutations in coding sequences*: Mutations in the ras gene have been detected in 30% of human cancers.
   ii) *Chromosomal aberrations*: It is believed that loss of tumour suppressor gene activity leads to malignancy, e.g. familial retinoblastoma—an autosomal trait; the absence of retinoblastoma (Rb) gene product in homozygous state leads to development of this tumour. Tumour suppressor activity of Rb gene has been demonstrated in vitro in cancer cells.

Retinoblastoma (Rb) gene: It encodes P105, a nuclear protein involved in regulation of cell cycle. It is also involved in transcription of critical genes.

p53 gene: p53 mutations are found in 50% of bladder, breast, colon and lung cancers. p53 has been recognised as "guardian of genome", since it allows DNA

Summary—cont'd

damaged through normal wear and tear to be repaired by preventing progress of the cell cycle.

Li-Fraumeni syndrome: It is a rare AD trait associated with point mutations in highly conserved regions of p53 gene (codons 245–258) involving germ cell line of family members. So the individuals are highly susceptible to variety of early-onset malignancies such as breast cancer, adrenal carcinoma, sarcoma, etc.

Deleted in Colorectal Cancer (DCC): DCC gene is expressed in normal colonic mucosa. It is reduced or absent in colorectal carcinoma. It is associated with 18q, loss of which is found in 70% colorectal cancers.

QUESTION YOURSELF*

1. What are cellular oncogenes?
2. What are viral oncogenes?
3. What are anti-oncogenes?
4. What are PDGF and EGF?
5. Which chromosomal aberrations are encountered in cancer?

*See page 265 for Answers.

11 Genetic Component in Common Diseases

LEARNING OBJECTIVES

At the end of this chapter the students should be able to understand

- Genetic susceptibility in population
- Familial occurrence of coronary artery disease, diabetes, hypertension, obesity, etc.

KEY WORDS

Genetic susceptibility

Conventionally, the focus in medical genetics has been on single gene disorders or unifactorial chromosomal disorders. Progress in medical genetics has certainly helped in identifying specific mutational error, thus offering better or rather more accurate risk estimate and subsequently more effective treatment. However, these conditions account for only a small fraction in total burden of human genetic diseases. Many common diseases such as diabetes, hypertension, stroke, coronary artery disease (CAD) and cancer have genetic components but are not caused by single gene or chromosome error. In fact, they are responsible for more morbidity and mortality in most countries. These diseases collectively warrant an urgent attention of the healthcare professionals. These diseases exhibit a complex interplay of genetic and environmental factors; in fact, it is rare for heredity or environment to be totally responsible for a particular disease. Mostly both factors operate in combination.

These diseases range between two ends—at one end we have a condition like Duchenne muscular dystrophy (DMD), which is entirely genetic in origin, while at the other end we have infectious diseases resulting from environmental factors. The "common diseases" referred above find place between these two extremes.

GENETIC SUSCEPTIBILITY

Some individuals are more prone to develop common diseases, like diabetes, CAD. This is because they have inherited predisposition, i.e. genetic susceptibility for that particular disease. Some of these diseases are because of single gene defect; however in majority of these diseases, this is due to effects of multiple genes, i.e. polygenic inheritance along with environmental factors.

Mechanism

Genetic susceptibility may be caused by single gene defect leading to abnormal gene product involved in a particular metabolic pathway, e.g. CAD in younger age group due to familial hypercholesterolaemia (FH). In this disease, the main cause is genetic susceptibility. This can, however, be mitigated by environmental factors such as:

1. Reducing dietary cholesterol
2. Avoiding smoking
3. Regular exercise
4. Avoiding obesity, etc.

In some diseases, environmental factors may be the main determinant, e.g. smoking in the development of pulmonary emphysema in individuals with α′1-antitrypsin deficiency. α′1-Antitrypsin is one of the main serum proteins. It inhibits activity of many proteolytic enzymes including trypsin. Individuals who are homozygous for *PiZZ*, the most common allele of this protease inhibitor, have greater risk of developing pulmonary emphysema and cirrhosis of liver. However, individuals who are heterozygous for α′1-antitrypsin deficiency are at increased risk of developing emphysema when exposed to environmental factors such as smoking or hazardous chemicals.

It would be interesting to note that in some instances, genetic susceptibility may be governed by single gene polymorphism and may vary in different populations, e.g. acetaldehyde dehydrogenase (ALDH) activity and alcoholism. In alcohol metabolism occurring in liver, enzymes involved include alcohol dehydrogenase (ADH) and ALDH. The human ADH consists of dimers with various combinations of subunits of three different polypeptide units coded by three loci. ADH1 coding for α subunit; ADH2 coding for β subunit and ADH3 that codes for γ subunit. In case of ALDH, there are two major variants—ALDH1 present in cytosol and ALDH2 present in mitochondria. The acute flushing reaction to alcohol in Far East Asians is said to be due to absence of ALDH2 activity. This unpleasant reaction to alcohol could account for lower incidence of alcoholism in this population. The genetic susceptibility can explain differences in responses to drugs, e.g. isoniazid inactivation status in the management of tuberculosis.

DEMONSTRATION OF GENETIC SUSCEPTIBILITY

Following approaches are available in order to determine genetic susceptibility:

1. Family study
2. Twin studies
3. Population studies
4. Immigration studies
5. Polymorphism associations
6. Biochemical studies
7. Animal models

Family Study: A higher frequency of family history of a disease in relatives than in general population is suggestive of genetic susceptibility. The proportion of the affected individuals in family as first- or second-degree relatives provides useful information for empiric recurrence risk and eventually helps in genetic counselling. Finding family aggregation of the disease, however, does not prove genetic susceptibility because the families have common environmental factors.

Twin Studies: If both the members of a pair of monozygotic (MZ) twins have the same trait, it does not mean that the trait is hereditary since they share same environment. The trait could be due to similar environmental factors to which they are exposed, e.g. a contagious disease like impetigo. This issue can be partly resolved by comparing findings between dizygotic (DZ) and MZ (identical) twin pairs.

The twins are said to be "concordant" when either both members are affected or neither is affected. The term "discordant" means only one member of the twin pair is affected. If a disease is entirely due to the environmental factors the concordance rates shall be very much the same whether they are identical or non-identical twins.

Population and Immigration Studies: The incidence of a particular disease is different in different population groups. This suggests the possibility of genetic factors being important. Immigration studies have suggested that an immigrant group moving from population group with low incidence of a disease to the one with high incidence shows rise in incidence. This means that environmental factor/s are operational; on the other hand, maintenance of low incidence of a disease in an immigrant group suggests that genetic factor/s are playing more significant role.

Polymorphism Associations: The human leukocyte antigens (HLA) of the major histocompatibility complex and specific disease associations have been found, e.g. insulin-dependent diabetes mellitus (diabetes type 1 or IDDM) and HLA class II alleles suggesting that it is an autoimmune disease; other diseases include rheumatoid arthritis, celiac disease and so on.

Animal Model: Experimental animals such as rats or mice are used. Progeny that has extreme values of a trait, e.g. rats having high blood pressure is selected. These are crossed with normal animals to produce offsprings. These offsprings have one normal chromosome and the other "affected" chromosome having gene causing high blood pressure. These animals are mated with normal animals. This produces third generation in whom one chromosome has normal genes and other chromosome has recombinations between normal and affected chromosomes (due to "crossing over" in parental meiosis). This series of mating produces progeny that is useful for linkage analysis.

High resolution genetic maps of these experimental animals are available. The polymorphic markers at regular intervals, ideally at every 10 cM, are then identified throughout the genome of the organism.

Linkage analysis is performed, comparing each polymorphic marker against the trait. Since animals with extreme values of the trait were selected, the procedure should reveal markers that are linked to loci that produce extreme phenotype. Once a linked marker is found, it is possible to isolate actual functional gene that is responsible for the trait.

When the functional gene is isolated and cloned, it can be used as a probe to search human genome for a gene with high DNA sequence homology that may have the same function. This is referred as "candidate gene". This approach is possible because the DNA sequences of functionally important genes in human and in experimental animals like rodents are often similar.

This approach is effectively used in studies relating to type I diabetes (IDDM) and hypertension. This technique detects only individual genes that cause disease in the animal model. However, in this approach one cannot assess the pattern of interactions of these genes, which may be critical and may differ in humans and animals. Nevertheless this helps us in identifying genes responsible for multifactorial diseases. In the following text, we shall deal with the genetic components involved in some of the common diseases.

Coronary Artery Disease

Heart disease is a major killer in most of the countries in the world. The most common underlying cause of heart disease is coronary artery disease (CAD). This is caused by atherosclerosis. It leads to narrowing of the coronary arteries, impairing blood flow to myocardium eventually resulting in myocardial infarction (MI). Numerous factors have been identified for CAD, such as obesity, smoking, hypertension, raised cholesterol level and, last but not the least, family history, which means having one or more affected first-degree relatives. Family studies have shown that an individual with positive family history is 2–7 times more likely to suffer from CAD than an individual with no family history of CAD.

The risk is higher:

(a) If there are more affected relatives
(b) If the affected relative is female
(c) If the age of onset of the affected relative is early, less than 55 years

It will be worthwhile to ascertain role of genes in familial clustering of CAD. Many investigators have studied the role of lipids in atherosclerosis with focus on determination of variation in circulating lipoprotein levels.

Isolation and cloning of the gene that encodes low-density lipoprotein (LDL) receptor has been a significant advance in this area. 1 in 500 individuals is heterozygous for a mutation in this gene. Such individuals have LDL cholesterol levels doubled, and the condition is referred to as familial hypercholesterolaemia (FH).

Familial Hypercholesterolaemia (FH)

It is transmitted as an autosomal dominant trait and is an important cause of heart disease accounting for nearly 5% of MIs in younger age group. About 1 in 500 persons is heterozygote. The plasma cholesterol levels in these persons are twice as high as normal, i.e. about 300–400 mg/dl. This results in accelerated atherosclerosis. These individuals present with *xanthomas,* i.e. distinctive cholesterol deposits in skin and tendons. Studies have shown that about 75% of men with FH develop CAD and about 50% suffer from nearly fatal MI. In accordance with Hardy–Weinberg law, approximately 1 in 1,000,000 births is homozygous for the FH gene. Homozygotes are more severely affected. Their cholesterol levels range from 600 to 1200 mg/dl. Most of them suffer MI before 20 years of age; youngest on record has been 18 months.

In order to understand the disease, we need to know little more about LDL receptors and cholesterol. All cells have cholesterol as a component of their plasma membrane. They either synthesise their own cholesterol or obtain it from extracellular environment, where it is carried primarily by low-density lipoprotein (LDL). By endocytosis, LDL-bound cholesterol is taken into the cell via LDL receptors along the cell surface. FH is basically due to reduction in number of functional LDL receptors on cell surfaces. Since these individuals lack number of LDL receptors, the cholesterol uptake is remarkably reduced which in turn increases circulating cholesterol levels.

LDL Receptors. In fact, the current knowledge that we have about endocytosis is based on LDL receptor study. The LDL receptors are glycoproteins; these are synthesised in endoplasmic reticulum. They then pass through Golgi and reach the cell surface. A part of the LDL receptor protrudes outside the cell.

The circulating LDL particle is bound to LDL receptor. LDL receptors lie in depressions called “coated pits”. These are coated with protein called Clathrin. The coated pit invaginates taking LDL particle inside the cell. Within the cell, the LDL particle is separated from the receptor and taken to lysosome. The lysosomal enzymes breakdown the LDL particle, while the receptor is recirculated to the cell surface. Each LDL receptor undergoes this cycle every 10 minutes. Free cholesterol is released from the lysosome for incorporation into the cell membranes, or metabolism into bile acids or sterols. Excess cholesterol is stored as cholesterol ester or removed from the cell by association with high density lipoprotein (HDL).

As the cholesterol levels in the cell rise, its synthesis is reduced by inhibition of an enzyme HMG-CoA reductase. The increment in cholesterol level promotes activity of acyl-coenzyme A: cholesterol acyltransferase (ACAT). This modifies cholesterol to be stored in the form of cholesterol esters. Simultaneously, the number of LDL receptors is decreased by reducing transcription rate of the LDL receptor gene (*LDL*R). Eventually, this reduces cholesterol uptake.

*LDL*R gene was cloned in 1984. This explains how receptor defects cause FH. *LDL*R gene is located on chromosome 19. It is 45 kb in length and has 18 exons and 17 introns. Over 900 mutations have been identified in this. About two-thirds of these are missense and nonsense substitutions. LDLR mutations can be grouped under five classes:

Class I: These mutations result in no detectable protein product. The heterozygotes produce only half the normal number of LDL receptors.

Class II: In this, LDL receptors are produced but cannot separate from the endoplasmic reticulum; eventually they are degraded.

Class III: Here LDL receptors are produced; they migrate to cell surface but fail to bind to LDL.

Class IV: These mutations produce LDL receptors that cannot migrate to specific "coated pits" and thus do not carry LDL into the cell.

Class V: These mutations produce LDL receptors that cannot dissociate from the LDL particle after entry into the cell. The receptor cannot return to the cell surface and finally gets degraded.

Each of the above stated mutations reduces the number of functional LDL receptors; this decreases LDL uptake, leading to rise in the circulating cholesterol levels. FH heterozygotes have half the normal number of LDL receptors, while homozygotes have no functional LDL receptors.

Management of FH

1. Dietary reduction of cholesterol with the reduced intake of saturated fats.
2. Administration of bile acid absorbing resins such as cholestyramine. It is interesting to note that reduced recirculation of cholesterol from the gut triggers the liver cells to form more LDL receptors, lowering circulating cholesterol levels. However, decrease in the intracellular cholesterol stimulates cholesterol synthesis in liver cells; thus, overall reduction in cholesterol level is 15%–20% only.
3. The above treatment is more effective when combined with "statin" group (e.g. lovastatin, pravastatin), which reduces cholesterol synthesis by inhibiting 3-hydroxy-3-methylglutaryl coenzyme-A (HMG-CoA) reductase. Decreased cholesterol level promotes formation of LDL receptors. In heterozygotes, this therapy often brings serum cholesterol levels to normal. However in homozygotes, it is less encouraging. *Liver transplants* with its limitations, because of availability of donors, seems to be the only answer. Hepatocytes of the transplant provide normal LDL receptors. Plasma exchange every 1–2 weeks brings cholesterol levels to 50%; however, it is difficult over long periods to practice this.

Somatic cell gene therapy with hepatocytes carrying normal LDL receptors is another option. These cells are introduced in portal circulation. This may be therapy in the future, which is being tested currently in FH homozygotes.

Another form of heart disease is cardiomyopathy, which involves cardiac musculature and leads to reduced cardiac function. *Hypertrophic cardiomyopathy* involving left ventricle is found in nearly 1 in 500 adults and accounts for about 10,000 deaths every year. Overall, 50% cases exhibit familial tendency. It is caused by autosomal dominant mutations in one of the ten genes that encode cardiac sarcomere. The most frequent being gene coding for β-myosin heavy chain, accounting for 35% of the familial cases; myosin-binding protein C for 20% cases; and troponin T for 15% of the cases. *Dilated cardiomyopathy* is seen with the frequency of 1 in 2500 individuals. About one-third cases show familial tendencies. Mostly it is because of autosomal dominant mutations; however, X-linked or mitochondrial inheritance is also known. The genes involved are those coding for cytoskeletal proteins such as actin, cardiac troponin T, components of the dystroglycan–sarcoglycan complex, desmin, etc.

Long QT (LQT) Syndrome

In this, the affected individuals have elongated QT interval in ECG. It is indicative of prolonged cardiac repolarisation. The disorder may be due to inherited mutations or caused by exposure to drugs that block potassium channels. This predisposes the affected individuals to potentially fatal cardiac arrhythmia.

Romano–Ward Syndrome

It is a potentially fatal cardiac arrhythmia. This can be caused by mutations involving any one of the five genes, among which four encode potassium channel subunits, they are KVLQT1, HERG, KCNE1 and KCNE2, while one encodes sodium channel and it is SCN5A.

Table 11.1 shows genes responsible for causing heart disease, their chromosomal association and also the corresponding protein product being affected.

Diabetes Mellitus

There are three main forms of diabetes mellitus (DM)—type 1 DM, type 2 DM and maturity onset diabetes of young. Clinically they are different in their presentations.

Type 1 DM

It is also called IDDM, i.e. insulin-dependent diabetes mellitus. It is seen in adolescence and can be controlled only by regular administration of insulin. It is characterised by T-cell infiltration of the pancreas leading to β-cell destruction. Apart from this, autoimmune antibodies against pancreatic cells are also observed. They appear even before clinical picture becomes evident.

Family Study: Sibs of individuals with type 1 diabetes have significantly high risk, approximately 6% as against 0.3%–0.5% in general population.

Table 11.1 Genes Contributing to Heart Disease

Gene	Chromosomal Location	Protein Product Function
Apolipoprotein A-I	11q	HDL component, LCAT factor
Apolipoprotein A-II	1p	HDL component
Apolipoprotein A-IV	11q	HDL and chylomicron component, HDL metabolism
Apolipoprotein B	2p	Involved in formation of LDL, VLDL, IDL and chylomicrons ligand for LDL receptor
Apolipoprotein C-I	19q	LCAT activation
Apolipoprotein C-II	19q	Lipoprotein lipase activation
Apolipoprotein D	2p	HDL component
Apolipoprotein E	19q	Ligand for LDL receptor
LDL receptor	19p	Uptake of circulating LDL particles
Lipoprotein (a)	6q	Cholesterol transport
Lipoprotein lipase	8p	Hydrolysis of lipoprotein lipids
LCAT	16q	Cholesterol esterification

HDL—High density lipoprotein; LDL—Low density lipoprotein; IDL—Intermediate density lipoprotein; VLDL—Very low density lipoprotein; LCAT—Lecithin–cholesterol acyltransferase.

(Adapted and Modified from King RA, Router JI. *Genetic Basis of Common Diseases*, 2 Edition, Oxford University Press, New York.)

The recurrence risk is also elevated, if the parent is affected. The risk for offspring of diabetic mothers is 1%–3%, while it is 4%–6% for offsprings of diabetic father.

Twin Study: Twin studies have shown empiric risk for MZ twins of type 1 DM to the tune of 30%–50%. In DZ twins, it is 5%–10%. The fact that type 1 diabetes is not 100% concordant in MZ twins signifies that genetic factors are not solely responsible for it. There is sufficient evidence indicating that some viral infections possibly have a role to play in causation of type 1 DM by activating an autoimmune response.

It is estimated that HLA system accounts for approximately 40% of the familial clustering of type 1 DM. About 95% of Caucasians with IDDM have HLA DR3 and/or DR4 alleles, while percentage of these alleles in general Caucasian population is about 50%. If an affected proband and a sibling are both heterozygous for DR3 and DR4 alleles, the risk of sibling developing type 1 DM is about 20%, i.e. approximately 40 times greater than in general population. It reflects over linkage disequilibrium between alleles of the DR locus and those of HLA-DQ locus. It is observed that absence of aspartic acid at position 57 of the DQ chain is associated with susceptibility to type 1 DM. The aspartic acid substitution alters the configuration of the HLA class II molecule and also its ability to bind and present peptides to T cells. This altered T-cell recognition may avert autoimmune episode in the individual.

Insulin gene located on the short arm of chromosome 11 is a logical candidate for type 1 DM susceptibility. Polymorphism studies involving this gene and region near it have been done. A strong association is seen with allelic variation in a variable number of tandem repeats (VNTRs) polymorphism located just at 5′ end of the insulin gene. The differences in number of VNTR units possibly affect transcription of the insulin gene, resulting in variation in susceptibility. Approximately, 10% of the familial clustering of type 1 DM is due to genetic variation in insulin region.

An animal model, using non-obese diabetic mouse, has been studied extensively. In addition, affected sib-pair analysis has also been used to map additional genes causing type 1 DM. At least 20 candidate regions have been identified that may harbour type 1 diabetes susceptibility genes. However, precise identification of these genes is a daunting task considering the complexity of the disorder such as locus heterogeneity, polygenic background and HLA alleles accounting for genetic susceptibility to type 1 diabetes.

Type 2 diabetes mellitus

Type 2 DM is also recognised as non–insulin-dependent diabetes mellitus (NIDDM). It accounts for more than 90% of diabetic patients. An endogenous insulin production is nearly always present in type 2 DM patients. The disorder is often effectively managed by dietary modifications and oral antidiabetic drugs. The disorder is typically seen in persons who have crossed 40 years of age, often associated with obesity. In fact, increasing obesity among younger age group in developed countries has shifted occurrence of type 2 DM to young adults. In contrast to type 1 DM, HLA associations and auto-antibodies do not seem to be operating in type 2 DM.

MZ twin concordance rates are substantially higher in this type of diabetes, often crossing 90%. Empiric recurrence risk for the first-degree relatives of the patient with type 2 DM is higher than type 1 DM patients, usually between 10% and 15%. Risk factors include two important ones: (*i*) family history and (*ii*) obesity—dietary habits and lack of exercise being responsible for it. Exercise reduces obesity, increases insulin sensitivity and eventually improves glucose tolerance.

An extensive linkage analysis undertaken to identify gene responsible for type 2 DM has led to a region on chromosome 2q. The gene encoding calpain-10, a cysteine protease, is associated with type 2 DM susceptibility. Another significant association of type 2 DM and a common allele of the gene that encodes peroxisome proliferator-activated receptor-γ (PPAR- γ) has been revealed. PPAR-γ is a transcription factor involved in adipocyte differentiation and glucose metabolism.

Maturity-onset diabetes of young (MODY)

This accounts for 1%–5% of the diabetic patients. It occurs typically before 25 years of age. It follows autosomal dominant inheritance. In contrast to type 2 DM, it is not associated with obesity. Family study reveals that about

50% cases result from mutations involving gene that encodes glucokinase (rate-limiting) enzyme in conversion of glucose to glucose-6-phosphate in pancreas. Mutations involving five other genes encoding for transcription factors in development of pancreas and insulin regulation have been identified as culprits in MODY. These are as under:

1. Hepatocyte nuclear factor-1 α′ (HNF1 α′)
2. Hepatocyte nuclear factor-1 β (HNF1 β)
3. Hepatocyte nuclear factor-4 α′ (HNF4 α′)
4. Insulin promoter factor-1 (IPF1) and
5. Neurogenic differentiation-1 (NEU ROD1)

Mutations involving these genes are expressed as dysfunction of pancreatic β cells leading to diabetes.

HYPERTENSION

The systemic hypertension is encountered in nearly 25% of the adult population in most of the countries in the world. In turn, it is responsible for heart disease, stroke and renal disease in many cases. Familial tendency is observed in about 20%–40% cases. Heritability estimates with twin studies is to the tune of 60%. The very fact that it is less than 100% in twin studies substantiates the role of environmental factors. More significant among the environmental factors are psychosocial stress, increased sodium intake, decreased physical activity and obesity.

Regulation of blood pressure is a highly complex process. It is governed by many physiological factors, such as cellular ion transport, vascular tone, heart function, kidney function, etc. Blood pressure variation is influenced by *rennin–angiotensin system, kallikrein–kinin system* (use of vasodilators such as nitric oxide) and ion-transport system e.g. *sodium–lithium counter transport.* Linkage and association studies have shown that gene encoding for angiotensinogen is responsible for causing hypertension.

Some of the hypertensive individuals present with single gene disorders, e.g. **Liddle syndrome.** These patients have low plasma aldosterone and hypertension. It is caused by mutations altering epithelial sodium channel (ENaC). In yet another syndrome called Gordon syndrome, hypertension is associated with high serum potassium level and increased renal salt resorption. It is caused by WNK1 or WNK4 Kinase gene mutations.

Nearly eight genes have so far been identified that can lead to some rare forms of hypertension. All of them affect resorption of water and salt by kidneys. This in turn has effect on blood volume as well as blood pressure. It may be expected that isolation and study of these genes may be helpful in identifying genetic factors involved in causing essential hypertension.

The biochemical studies of hypertensive individuals have supported possibility of a common autosomal dominant gene. In some hypertensive individuals, there is defective extrusion of sodium from red cells because of defect in enzyme that controls sodium, potassium cotransport at red cell membrane. This results in accumulation of sodium within the cell, and the increased intracellular sodium is major determinant of hypertension.

Another red cell ion transport assay has been developed. It is called sodium–lithium counter transport system. It involves loading RBCs with lithium and measuring the rate of efflux of lithium from RBCs in low- and high-sodium concentrations. An increased sodium–lithium counter transport correlates with hypertension.

Yet another assay called ouabain-sensitive sodium–potassium ATPase system has been contemplated as a possible indicator of genetic susceptibility. The above stated assays provide supportive evidence for the role of single gene in determining predisposition to essential hypertension.

Other Factors: Women who develop pre-eclampsia in gestation period were thought to possess an increased risk of hypertension following the pregnancy. Pre-eclampsia is a condition presenting with oedema, proteinuria along with hypertension. A careful review and assessment of these cases, however, revealed that these women already had pre-existing predisposition to hypertension that is incidentally detected during pregnancy.

Obesity

Obesity is defined as body mass index greater than 30.

$$\text{BMI} = \frac{W}{H^2} \text{ i.e. } \frac{\text{Weight in kilograms}}{(\text{Height in meters})^2}$$

With this definition, nearly 30% adult population in developed countries is obese. About 35% individuals are overweight (BMI ranging between 25 and 30). Though obesity is not a disease, it certainly is an important risk factor for number of common diseases, e.g. stroke, type 2 diabetes, hypertension, etc. With the current lifestyle in the society, obesity is on rise. There is strong correlation between obesity in parents and in their children. This may be logically ascribed to the common environment that they share with similar dietary and exercise habits. However, there is sufficient evidence suggestive of genetic component also being responsible. This is attested by adoption studies where the body weights of adopted individuals correlated with their biological parents and not with their adoptive parents.

Recently, a study on mouse models has shown several genes responsible for obesity. More important being genes encoding for "Leptin", a hormone, and its receptor. Leptin is secreted by adipocytes (fat cells) and acts on appetite control centre in the hypothalamus. Increased fat stores cause elevated leptin level; this stimulates satiety centre and leads to loss of appetite. Low leptin levels increase appetite. Mice with mutations involving leptin gene or leptin receptor gene develop obesity.

Human homologs of leptin gene and its receptor, however, do not support the findings completely. Most obese humans show high leptin levels indicating that leptin gene is functioning normally; with this, the leptin receptor gene was suspected to be defective. Individuals with severe obesity (BMI more than 40) show mutations involving leptin gene and its receptor gene.

Additionally, leptin participates in important interactions with other components that have an effect on appetite, such as neuropeptide γ and α'-melanocyte stimulating hormone and its receptor, i.e. melanocortin-4-receptor (MC4R). In fact, 3%–5% of the severely obese individuals have mutations involving MC4R. Learning more about human genes responsible for obesity would eventually lead us to better and effective management of the problem.

PEPTIC ULCER

It may be in the stomach (gastric) or in the duodenum (duodenal ulcer). This can be ascertained on gastroscopy or on radiological examination. Past studies have shown that gastric ulcer is more frequent in poor socioeconomic group, while duodenal ulceration is more frequently found in an affluent group, i.e. higher socioeconomic group. This suggests that there are different environmental factors involved. Numerous environmental factors seem to have their role including smoking, stress, alcohol consumption, etc., in peptic ulceration. Another factor, an infective agent, *Helicobacter pylori* is also said to cause chronic gastritis leading to peptic ulceration.

Past studies have indicated that peptic ulceration is twice as common in males as in females; however, recently the sex ratio has nearly been equal. The underlying cause for this change might be smoking and work-related stress in women in recent times.

Family Study and Twin Study. Peptic ulceration (gastric or duodenal) is twice as common in first-degree relatives of an affected person as in general population. It is suggested that this could be because of common environment that they share. Twin studies of peptic ulceration have shown concordance rate in MZ twins to be double as compared to the DZ twins. The MZ twins do not show 100% concordance; this indicates role of environmental factors in the aetiology of peptic ulcer.

Blood Group Association

Numerous studies conducted in different population groups have shown increased proportion of people with peptic ulceration to have "O" blood group. This does not mean that all O group individuals shall have or develop peptic ulcer. However, one can say that "O" blood group individuals are at higher risk of having peptic ulcer, to the tune of 30% as compared to other blood group, i.e. A, B or AB. Among the two, the duodenal ulceration is more closely (and frequently) associated with "O" blood group than gastric ulcer. Finding correlation with the secretor status and the peptic ulceration, it is revealed that peptic ulceration is more common in non-secretors. The non-secretor being 50% more likely to develop peptic ulceration. The combination of non-secretor status with "O" blood group has 2.5 times the risk of developing peptic ulceration as compared to general population.

Inheritance. Peptic ulcer can occur as a feature of some rare single gene disorder, such as multiple endocrine neoplasia type 1. It is an autosomal dominant trait with the risk of developing adenomas involving pituitary, parathyroid, pancreas with associated peptic ulceration.

ALZHEIMER DISEASE

Alzheimer disease (AD) is responsible for 60%–70% cases of progressive cognitive impairment in elderly people. It affects nearly 10% of population above 65 years and 40% of individuals above 85 years of age. It is characterised by progressive dementia and formation of amyloid plaques and neurofibrillary tangles in cerebral cortex and hippocampus. These changes lead to neuronal loss.

The risk of developing AD doubles in persons who have an affected first-degree relative. It follows an autosomal dominant inheritance in about 10% cases; though most cases do not appear to be caused by single loci. It is genetically heterogeneous disorder. Nearly half of early onset cases can be attributed to mutations in any of the three genes that affect amyloid-β deposition. Two of these are presenilin 1 (PS1) and presenilin 2 (PS2). They are similar to each other, and their protein product helps cleavage of the amyloid- β-precursor protein (APP). When APP is not normally cleaved, a long form of it accumulates and is deposited in the brain. This is primarily the cause of AD. Mutations involving PS1 gene, which is said to cause early onset of AD, may be in the fifth decade. Some cases present with mutations involving gene encoding for APP, located on chromosome 21. These mutations disrupt normal secretase cleavage sites in APP leading to accumulation of longer protein product. Since this gene is located on chromosome 21, some Down syndrome patients with an extra gene show amyloid deposits in brain and occurrence of AD.

For late-onset AD, an important risk factor identified has been an allelic variation in apolipoprotein E (APOE) locus. It has three major alleles—ϵ2, ϵ3 and ϵ4. Persons who have one copy of ϵ4 (heterozygous) are at least 2–5 times more likely to develop AD, while homozygous individuals having two copies of ϵ4 are 5–10 times more likely to develop the disease. Difference exists among populations—Europeans and Japanese have higher ϵ4 associated risk, while Hispanics and African–Americans have relatively lower risk.

Genome scans indicate that there are additional AD genes on chromosomes 10 and 12. A gene located on chromosome 12 region encodes α_2-macroglobulin. It is protease inhibitor that interacts with apolipoprotein E. Another gene in this region codes for low-density lipoprotein receptor-related protein (LRP), which too interacts with apolipoprotein E. Some studies have supported an association between these alleles and late-onset AD. Many features associated with Alzheimer pose difficulty in genetic analysis of the disease. These are as under:

1. Genetic heterogeneity.
2. Difficulty in definitive diagnosis. It is only possible with brain autopsy. Diagnosis in living members is indicated through clinical features and brain imaging studies.
3. Since the disease occurs late in life, the individuals carrying AD-predisposing mutations could die from another cause before AD is fully manifested. These individuals are likely to be missed as non-carriers.

Despite all these hurdles, several genes associated with AD have been identified, leading to better understanding and eventually more effective treatment of the disease.

Summary

Many common diseases such as diabetes, hypertension, stroke, coronary artery disease and cancer have genetic components involved, but are not caused by single gene or chromosome error. They cause more morbidity and mortality than the genetic diseases and this fact warrants urgent attention of healthcare professionals.

Genetic Susceptibility: Some people are more prone to develop diseases like diabetes and hypertension; it means, they have inherited predisposition. This is called genetic susceptibility.

Mechanism: Caused by single gene defect leading to abnormal product causing metabolic error.

Genetic susceptibility can be determined through following: (*i*) Family study; (*ii*) Twin study; (*iii*) Population study; (*iv*) Immigration study; (*v*) Biochemical study; (*vi*) Polymorphism associations; (*vii*) Animal models.

Coronary Artery Disease (CAD): A major killer world over. Cause can be familial hypercholesterolaemia (FH).

Familial hypercholesterolaemia (FH): It is an autosomal dominant trait. Affects 5% of younger age group myocardial infarctions (MI). Patients have double plasma cholesterol level than normal. This leads accelerated atherosclerosis. 75% of men with FH develop CAD, and 50% suffer massive MI. One in 1000,000 births is homozygous for FH gene.

Low density lipoprotein (LDL) binds with cholesterol and is taken within the cell by LDL receptors. In FH, number of functional LDL receptors is reduced; this reduces cholesterol uptake in the cell, in turn increasing the circulating cholesterol levels. This promotes atherosclerosis. LDL receptor is a glycoprotein, synthesised by endoplasmic reticulum. LDL receptor gene is located on chromosome 19. It is 45 Kb long and has 18 exons and 17 introns. There are five classes of mutations involving it—classes I–V.

Management: Dietary reduction of cholesterol, administration of bile acid absorbing resins, e.g. cholestyramine. Use of statin group drugs (e.g. lovastatin, pravastatin) helps effectively. *Somatic cell gene therapy* with hepatocytes carrying LDL receptors may also help.

Continued

Summary—cont'd

LQT Syndrome: Mutations or use of potassium channel blocking drugs cause long QT interval in ECG, which may cause fatal cardiac arrhythmia.

Romano–Ward syndrome: A potentially fatal cardiac arrhythmia caused by mutation of genes encoding potassium channel subunits.

Diabetes Mellitus (DM)

Type 1 DM: Also called insulin-dependent diabetes mellitus (IDDM). It can be controlled only by insulin. It is caused by destruction of β cells. Autoimmune antibodies cause this. Viral infections possibly play a role in causation of IDDM.

About 40% of familial clustering of type I DM is due to HLA system. Absence of aspartic acid at 57 position of DQ chain at HLA–DQ locus is the cause. In 10% cases, alterations in insulin gene at 11p is responsible for familial clustering of type 1 DM.

Type 2 DM: It is non–insulin-dependent diabetes mellitus (NIDDM); seen in age group above 40 years. Gene encoding for Calpain-10, a cysteine protease, located on 2q is associated with type 2 DM.

Maturity-Onset Diabetes of Young (MODY): It is typically seen before 25 years of age; it is an autosomal dominant trait and is not associated with obesity. Mutations of gene encoding glucokinase enzyme and also five genes encoding for transcription factors in development of pancreas and insulin are probable culprits.

Hypertension: It is seen in 25% of world population. It may cause heart disease, stroke, and renal disease. Factors responsible include familial tendency (in 20%–40%), obesity, more of sodium intake, stress, etc. Physiological factors include cellular ion transport, vascular tone, heart and kidneys function.

Liddle syndrome: Single-gene disorder with low plasma aldosterone and hypertension. In *Gordon syndrome,* hypertension is associated with high serum potassium and is due to mutations of *WNK*1 or *WNK*4 kinase gene. Increased intracellular sodium is the major determinant of hypertension. Red cell ion transport assay and another assay involving ouabain-sensitive sodium–potassium ATPase system are indicators of genetic susceptibility.

Obesity: With body mass index (BMI) more than 30, one can put label of obesity. Mouse model studies have shown genes encoding leptin, a hormone, and its receptor associated with adipocytes being responsible for this. Interactions of leptin with other components (e.g. neuropeptides γ, α′-melanocyte stimulating hormone and its receptor [i.e. MC4R]) govern appetite, and the mutations of these may cause obesity.

Peptic Ulcer: It may be in stomach (known as gastric ulcer) or duodenum (known as duodenal ulcer). Gastric ulcer is seen in poor socio-economic group, while duodenal ulcer in affluent group. Peptic ulcer is twice as common in males as in females. Factors responsible are stress, smoking, alcohol consumption or *H. pylori* infection. O blood group individuals are more among peptic ulcer patients.

Alzheimer Disease: It presents with progressive dementia, amyloid plaques, and neurofibrillary tangles in cerebral cortex and hippocampus with neuronal loss. It is seen in 10% of people above 65 years and 40% individuals above 85 years. It is an autosomal dominant trait. Mutations in any of the three genes that affect amyloid-β deposition (i.e. presenilin 1 [PS1], presenilin 2 [PS2] or gene encoding amyloid- β precursor protein [APP]) account for 50% of early-onset cases.

Late-onset Alzheimer cases show allelic variation in apolipoprotein E (APOE) locus. Genome scans have indicated Alzheimer disease gene on chromosome 10 and 12. The latter encodes α_2-macroglobulin.

QUESTION YOURSELF*

1. What do you mean by genetic susceptibility?
2. What causes genetic susceptibility?
3. How can genetic susceptibility be determined?
4. What are LDL receptors?
5. Where is the locus for LDL receptor gene (LDLR)?
6. What does IDDM stand for?
7. What is body mass index?

---

*See page 266 for Answers.

12 Population Genetics

COMPETENCY 80.4 Describe the embryological basis of twinning in monozygotic and dizygotic twins.

LEARNING OBJECTIVES

At the end of this chapter the students should be able to understand

- Principle of Hardy–Weinberg's law
- Factors influencing Hardy–Weinberg's equilibrium
- Importance of Twin study
- Dermatoglyphics its principles and application

KEY WORDS

Beanbag genetics, Eugenics, Cloning, Dermatoglyphics

Population genetics deals with the study of genes in population. It also tells us about how distribution of genes and the genotypes are maintained or changed in population. Since both environment and heredity are operational in population genetics, a new field called *genetic epidemiology* has emerged. It deals with the distribution, aetiology and course of some heritable diseases in population. If we assume that the gene frequencies in population do not change at all, there will be no place for evolution. This means, the genetic profile of a population group does not always remain constant but changes over generations, offering a chance of evolution. The population genetics restricts itself to the study of one species, while evolution encompasses many of them. Population genetics and evolution are together designated as evolutionary genetics. Almost three billion years ago what existed as a cluster of precellular organic molecules has become a multicellular organism of today's human being.

HARDY–WEINBERG'S LAW

Hardy and Weinberg in 1908 independently defined this law. It is named after GH Hardy, an English mathematician, and W Weinberg, a German physician.

The similarity between their works was recognised in 1943 by C Stern who suggested the names of Hardy and Weinberg be given to this principle in population genetics. The law states, "gene frequencies in a population remain constant from generation to generation if no evolutionary factors such as migration, mutation, selection and drift are operating". The law provides a simple algebraic formula to calculate expected gene and genotype frequencies in population.

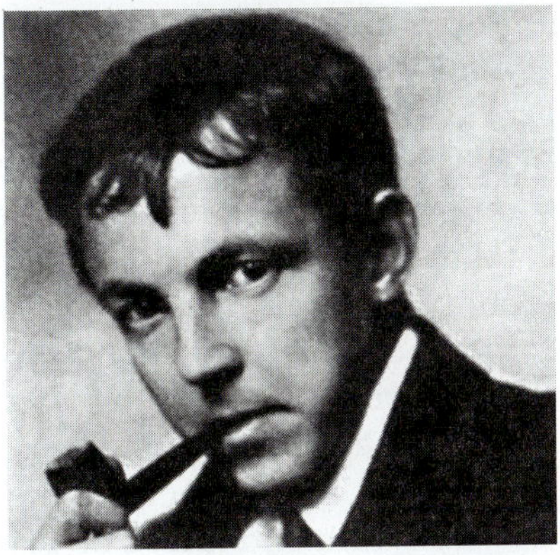

GH Hardy

(*Source: Historia Mathematica*, Adrian C. Rice and Robin J. Wilson, Volume 30, Issue 2, Pages 173-194, Elsevier, May, 2003.)

W Weinberg

(*Source: Principles of Behavioral Genetics:* Robert R.H. Anholt and Trudy F.C. Mackay, Figure 5.2, Page 72, Elsevier Inc., 2010.)

BEANBAG GENETICS

Population genetics is sometimes called beanbag genetics (Fig. 12.1). Let us consider a beanbag containing beans of two colours, black and white. We assume a system of only two alleles, being represented by the black and the white beans. The total beans in the bag constitute a gene pool. Among them assume frequency of black beans as p and that of white beans as q. If one draws two beans every time from this bag, three possible combinations can be expected—2 black; 1 black, 1 white; and 2 white. Considering p and q frequencies to these beans, the relative proportions of three combinations are as under:

$$p^2 \text{ (2 black)} + 2pq \text{ (1 black, 1 white)} + q^2 \text{(2 white)}.$$

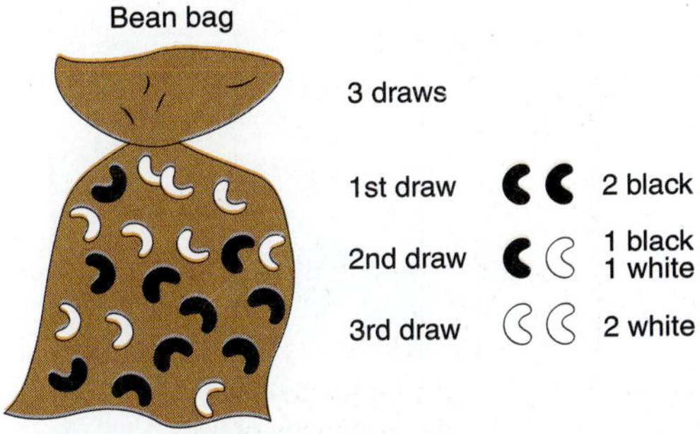

Figure 12.1 Concept of "beanbag genetics".

CALCULATING GENE FREQUENCY

We assume a population group of 1000 persons, among which there are individuals of the following blood groups: A group—200, AB group—500 and B group—300. These blood groups are determined by two alleles I^A and I^B. With two alleles A and B at AB locus in this population of 1000 persons, the total genes in the gene pool are $1000 \times 2 = 2000$ genes. The genotypes in this population will be AA—200, AB—500 and BB—300. Total number of allele A is: (AA) 200 + 200 + (AB) 500 = 900A. Therefore, frequency of allele A is 900/2000 = 0.45. In the same manner, the total number of allele B is: (BB) 300 + 300 + (AB) 500 = 1100B and its frequency as 1100/2000 = 0.55. Gene frequencies are expressed as fractions unity i.e. 1. Assuming frequencies as p and q, $p + q = 1$. Substituting the values in our example, $p\ (0.45) + q\ (0.55) = 1$. Genotype frequencies can also be obtained with the help of the algebraic formula derived earlier—$p^2 + 2pq + q^2$.

If we consider a system of two alleles M and m with frequencies p and q, respectively, there is a possibility of genotypes, MM, Mm and mm. The genes M and m occur with the same frequency in sperms and ova. The frequencies of offsprings from such mating would be p^2 (MM), $2pq$ (Mm) and q^2 (mm) (Fig. 12.2). If these progeny were to mate with each other, the resultant would be as depicted in Table 12.1.

If we sum up various types of offsprings:

Total of MM offsprings: $p^4 + 2p^3q + p^2q^2 = p^2\ (p^2 + 2pq + q^2)$
Total of Mm offsprings: $2p^3q + 4p^2q^2 + 2pq^3 = 2pq\ (p^2 + 2pq + q^2)$
Total of mm offsprings: $p^2q^2 + 2pq^3 + q^4 = q^2\ (p^2 + 2pq + q^2)$

Eliminating the common factor $(p^2 + 2pq + q^2)$, the proportions of genotypes MM, Mm and mm appear to be the same, i.e., $p^2{:}2pq{:}q^2$ as in the previous generation. The result would be the same if we continue to calculate for many generations. This clearly indicates that the gene and genotype frequencies are

Ova \ Sperms	M(p)	m(q)
M(p)	MM (p^2)	Mm (pq)
m(q)	Mm (pq)	MM (q^2)

p = frequency of "M"
q = frequency of "m"

Figure 12.2 Genetic combination in the two allele system of "M" and "m". Gene frequencies are given in parenthesis.

Table 12.1 The Frequency of Mating Types and Offspring

Mating Type	Frequency	Offsprings		
		MM	Mm	mm
MM × MM	p^4	p^4	–	–
MM × Mm	$4p^3q$	$2p^3q$	$2p^3q$	–
Mm × Mm	$4p^2q^2$	p^2q^2	$2p^2q^2$	p^2q^2
MM × mm	$2p^2q^2$	–	$2p^2q^2$	–
Mm × mm	$4pq^3$	–	$2pq^3$	$2pq^3$
mm × mm	q^4	–	–	q^4

maintained from generation to generation. With the help of this formula, one can calculate frequencies of various genotypes if the frequency of one of the homozygote is known. If the frequency of an autosomal trait is 1 in 10,000, i.e.

$$q^2 = 1/10{,}000,\ q = 1/100 \text{ and } p = 1 - 1/100 = 99/100$$

The frequency of carrier, i.e. $2pq = 2 \times 99/100 \times 1/100 = 1/50$

From this example, we learn that the frequency of carriers in population is much more than affected individuals.

Factors Influencing Hardy–Weinberg's Equilibrium

The Hardy–Weinberg's equilibrium holds true for large populations with a constant rate of mutation, random mating and no migration.

Mutation

Mutation usually causes loss or change of function of a gene. The spontaneous mutations occur with a frequency ranging from 1 in 10,000 to 1,000,000 per locus per generation. An average rate of mutation is about 1 in 100,000. Mutation rate can be calculated by the direct method for an autosomal dominant trait. However, for an autosomal recessive or an X-linked trait, an indirect method has to be adopted for estimation of mutation rate.

Direct Method

Let us consider an example of achondroplasia. Assume that out of 96,000 births in a hospital, 10 are achondroplastics. Two of these 10 children have affected parents, i.e. 10 – 2 = 8 are affected because of new mutation. This means, there is one case of mutation approximately in 12,000 births. In an affected child, mutation might have occurred in the gene from either the maternal or paternal side. This amounts to mutation rate per gene as 1 in 24,000, i.e. 42 mutations per million genes per generation.

Mutation Rate by Indirect Method

Let us assume μ as mutation rate per gene per generation. For an autosomal recessive trait, the formula is $\mu = F(1 - f)$ and for X-linked recessive trait, it is $\mu = 1/3F(1 - f)$. Here, F is the frequency of condition and f stands for reproductive fitness. The mutation rate is increased by a number of agents that include radiation (X-rays) and some chemicals like mustard gas or acridine orange.

Mating patterns

Random Mating: In this, there is no preferential selection of a mate of a particular genotype. However, this is not true in the strict sense of the word, e.g. ABO blood group. An individual's blood group (A or B) would hardly influence his choice of partner but this is untrue for other traits such as stature.

Non-Random or Assortative Mating: This type of mating pattern involves preferential selection of a genotype. Assortative mating increases the proportion of homozygotes in the population. Inbreeding has a similar effect in small populations often called as isolates. Members of such a group are prevented from marrying outside the group on cultural or religious grounds.

Random genetic drift

This involves variation in the number of children produced by individuals having different genotypes. This does not affect gene frequencies in large populations; in small isolated populations, this alters gene frequencies and disturbs the Hardy–Weinberg's equilibrium.

Migration

Mass migration of people into new territories would bring them in contact with diverse populations resulting in an exchange of genes between two groups. This is called *gene flow*. For example, the frequency of the gene responsible for B blood group is above 25% in Asiatic countries; however as we move westward, it decreases. In Scandinavia, it is less than 10%. This has been explained by the migration of Mongoloids towards the west from 500 AD to 1500 AD.

EUGENICS

It is derived from Greek word meaning "good birth". The eugenic movement has been initiated in the early part of last century, in Europe and United States of America. In the initial phase there were two entities, "positive eugenics", i.e. preferential reproduction of those deemed to genetically more fit, and "negative eugenics", i.e. preventing reproduction of those who are genetically weak/less fit. However, with this concept of negative eugenics Nazi Germany has written Black chapter in the human history. This reminds us of misuse of genetic information. Eugenics is a branch of genetics that promotes the improvement of hereditary qualities of a race/species.

Actually, the concept of eugenics was introduced by F Galton in 1883. It refers to establishing best characters in population by selective breeding. Eugenic movement in the plant kingdom and in animals has gathered momentum and has certainly been rewarded in the last few decades. It has given us a high yield of food grains and an excellent breed of animals. However, in human beings that cannot be considered as an experimental model for numerous reasons, eugenics cannot be effectively conceived. There are social, cultural and ethical problems challenging the human eugenic movement. A positive role of eugenics can be thought over by following examples:

1. Enacting a law under which mentally retarded persons must undergo compulsory sterilisation.
2. Promoting intelligence whereby Nobel laureates become donors in sperm banks.

But there are a few problems in the eugenic movement, for example:

1. Selection of characters to be promoted in the population. It is often difficult to decide on such issues.
2. Selective breeding essentially reduces genetic diversity.

Eugenic implications are indicated through activities carried today under genetics. This includes prenatal diagnosis of a genetic disorder. If diagnosis of a suspected disorder is made, it is followed by termination of pregnancy thereby preventing an addition of a handicapped individual in the population. Artificial insemination also forms a step in eugenics.

TWINS

AN 80.4

Twins are the commonest form of multiple pregnancy. They hold a special place in genetics because of their utility in comparing effects of genes and environment. The frequency of multiple pregnancy found today is possibly because of the frequent usage of human gonadotropins in cases of ovulatory failure. Another contributing factor being in vitro fertilisation (IVF) leading to multiple pregnancy. The incidence of twinning is around 1 in 90. If both members of a twin pair exhibit a trait, they are said to be concordant. In disorders of genetic aetiology, monozygotic twins show a higher degree of concordance as compared to dizygotic twins. In case monozygotic twins are not fully concordant in a particular condition, there must be non-genetic factors playing a role in the aetiology of the condition.

Types of Twins

AN 80.4

Twins are of two types:

1. Monozygotic
2. Dizygotic

Monozygotic twins or identical twins

They develop from a single zygote, which divides in the early embryonic life. Foetal membranes vary depending upon the time of twinning. If division of inner cell mass occurs after formation of amniotic cavity, i.e. after 8 days, then the monozygotic twins shall have one amnion (monoamniotic) and one chorion (monochorionic). If the separation of embryonic primordium occurs before development of amnion, then there are two amnion, two chorion and two placentae. This may pose difficulty in determining the twin zygosity. Since monozygotic twins result from a single zygote, they are always of the same sex. They are genetically identical and are alike in their genetic markers. Dissimilarity between monozygotic twins for certain traits like birth weight or size is influenced by environment, e.g. prenatal nutrition.

Dizygotic twins

Dizygotic twins account for two-thirds of twins. These twins result from fertilisation of two ova shed at the same menstrual cycle by two separate sperms. Genetically, they are no more than brothers and sisters born at different times. Dizygotic twins have an average half of their genes in common. Tendency of dizygotic twins repeats in family. With the first twin birth, the tendency of multiple births in subsequent pregnancies is almost five times more common than in the general population.

Determination of Twin Zygosity: Information about zygosity of a twin pair helps in the case of a genetic disorder or in transplantation. For example, in maturity onset type of diabetes, concordance in monozygotic twins is almost 90%. Twin zygosity can be determined by an examination of foetal membranes and placenta. Other characters such as eye colour, finger prints also help in the determination of twin zygosity.

ABNORMAL TWINNING

AN 80.4

Conjoined Twins

They arise from an incomplete separation of inner cell mass or embryonic disc. They are classified depending upon the part of body which is attached, e.g. thoracopagus. This indicates a union of thoracic regions. About 1 in 40 monozygotic twins do not separate completely and form conjoined twins (Fig. 12.3).

Chimaera

They are the individuals having cells derived from two different zygotes. Chimaera can be of the following types:

1. Dizygotic twins with exchange of blood cells between the twin members
2. Dispermic chimaera
3. Experimental chimaera (for details refer Chapter 4.)

Monozygotic Twins with Different Karyotypes

This forms another type of abnormal twinning. They possibly result from postzygotic non-disjunction followed by twinning.

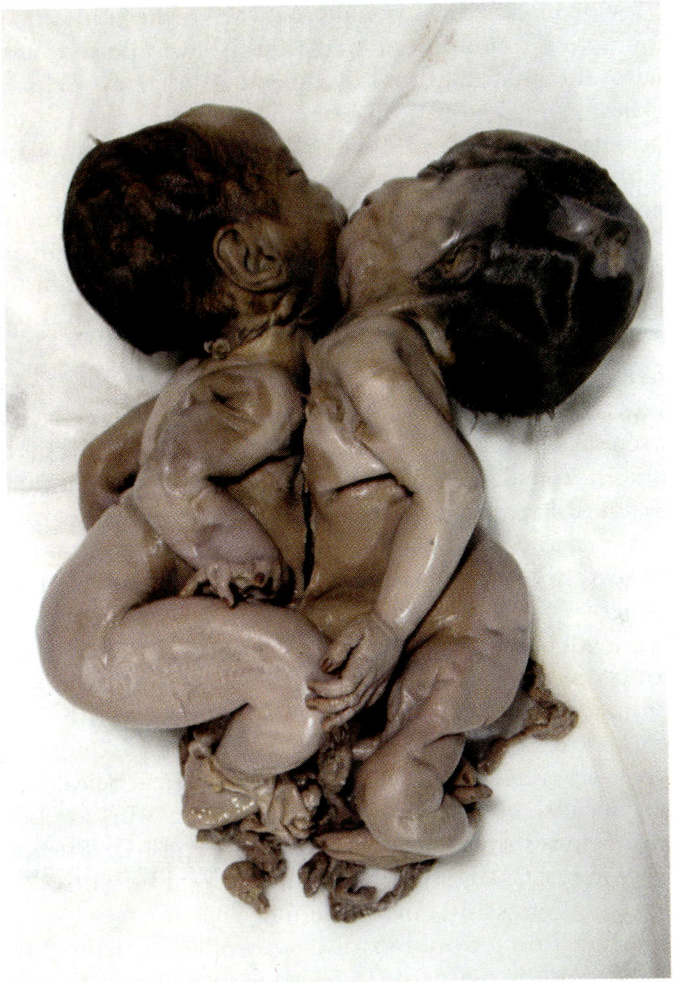

Figure 12.3 Conjoined twins showing thoracopagus.

CLONING

AN 80.4

"Dolly", the first life created in 1997. She is the first clone. A life blown from a single cell. She makes a proud marvel of modern medicine. The project was engineered by a group of geneticist from England. Scientists have deciphered a blue print of human genome; the time is not far, when human cloning would be possible.

If at all one thinks of human clone let us know what is the aftermath. US house representatives grappled for more than 3 hours with moral and legal aspects of human cloning before voting 265 to 162 votes to approve the "Human Cloning Prohibition Act of 2001". There is provision of steep criminal and civil penalties on any individual accused under the act. Participation in human cloning in anyway, ranging from creating cloned human cells to receiving medicine based on such research, may bring a 10-year prison term and if done for profit, the civil penalty may be to the tune of $1 million. A narrower, competing amendment that would have allowed cloning for research was also defeated, 249 to 178 votes. In short, the White House has strongly backed a complete ban on human cloning.

Japanese scientists claimed in 1998 to have cloned eight calves from the cells of a single adult cow, using the same technique that was used by Scottish scientists to develop Dolly. They transferred the nuclei from cells removed from a single adult animal into cow eggs from which the nuclei were already removed. The eggs bearing the transferred cell nuclei were grown under optimum conditions, into blastocysts. Ten blastocysts were placed into five unrelated cows. Total eight calves were born from these 10 blastocysts. Out of these eight, four died soon after birth due to environmental factors, as stated by the researchers.

Dr Mark West Husin, Texas-based livestock reproduction researcher quotes that more research is required to make it cost effective. The current status of cattle cloning, where half the calves die despite their birth under best conditions, is far from desired. The technique eventually will be important to the industry for its milk and meat production. Dolly's creators also had hundreds of failures. Researchers reported a success rate of only 12%. This means that there is lot of scope for improvement.

Yet another myth about cloning is that "cloning is creation of a carbon copy". Do not be excited at the idea of having your duplicate. A US research company "Clonaid" puts it simply as a remake plus or minus few genes. "An identical twin is closer to your personality than a clone", quotes Dr Mitradas Panicker, a researcher at National Centre for Biological Sciences, Bangalore. It is essentially because identical twins (monozygotic twin) are born from the same egg, while clones originate from different eggs. Dr Pushpa Bhargava, Founder Director of CCMB, Hyderabad, puts little differently, "A person is a consequence of his genes and environment". This means, to create a perfect mirror image, the cloning would involve controlled environment too.

Cloning blanks out the male role in reproduction; however, this could lead to genetic snafus. With the menace of genetic distortions, some scientists have objected to cloning. Cloning poses an interference with evolutionary chain. Clones would deprive us of the evolutionary uniqueness.

Proponents of cloning contend that the technology cannot be ignored considering its use in organ transplants. Organ transplants were thought to be sacrilegious few years ago. Pig organ transplants are now being contemplated for human patients in the UK.

The advances in genetic engineering cannot be summarily rejected with the threat of mishap. It will take some time to get accepted as a way of life. A drug costing $ 10,000 per gram, if genetically engineered through animals could cost $ 1 per gram, for example, factor VIII, which is used to treat blood disorders, is an expensive drug. The cost of such drugs can be reduced and be brought within reach of a common man.

Cloning could be an option for childless couple who do not wish to use donor sperm. However, it would not be truly their child but only a shadow of one of the parents. All the researchers in the field are unanimous on the need to test the cloning technology very thoroughly in animal settings before extrapolating it for human asexual replication. A breakthrough is no doubt exciting but should not blind us of humanistic aspect.

Possible Use of Cloning and Stem Cells in Curing Diseases

Heart disease

Among the common diseases, heart diseases are probably on the top. It kills more Americans than anything else, causing 41% of all deaths. Do not worry, now we have some **"Hi-tech Solutions for Heart Ailments".** There was a scientific session organised by American Heart Association—latest updates on using gene therapy to help people "grow their own heart bypasses". This, however, remains an experimental approach. It is based on a protein called Vascular Endothelial Growth Factor (VEGF). It helps formation of new blood vessels. Production of this protein is governed by a gene. Researchers use both the gene and its protein product in critically ill heart disease patients who cannot have anymore bypasses. Preliminary report has suggested that with this approach the patients have improved, their chest pain has reduced and these patients have returned to near normal activities.

Another promising approach has been with the help of stem cells or master cells. Canadian scientists have tried this in animal models. The bone marrow stem cells are used to grow into a new tissue to replace dead myocardium. If it works in humans, it may help numerous people who have damaged myocardial tissue, following heart attacks.

Diabetes

Manufacturers of Dolly (sheep) have made an announcement that it is aiming at genetic cure for diabetes—a slow silent killer. This brings a new hope to millions of diabetic patients. It afflicts Indians and South Asians more than white people. A project worth about 2 million dollars aims at development of pluripotent cells from non-human primates and livestock species. This may very well be of use in the management of diabetes and other conditions like Alzheimer disease; Parkinson disease; spinal cord injuries; burns; heart, liver and lung diseases; and cancer.

DERMATOGLYPHICS

The word dermatoglyphics means writing on the skin. This includes ridge patterns on the skin of palms, digits and soles. Its application in genetics is chiefly because of its diagnostic value. The dermatoglyphic patterns in some genetic disorders are characteristic and to some extent they also help in determining twin zygosity. The ridge patterns on hands and feet start developing around the 13th week of gestation and are completed by about the 16th week. Afterwards, these dermal patterns remain permanent. The scientific basis of dermatoglyphics was laid down by Galton much earlier. It was in 1961 that Cummins introduced the term "dermatoglyphics".

According to Galton's system, fingerprints can be classified into three basic patterns—arches, loops and whorls. This classification is based on a number of triradii. Arch has no triradius, loop has one and whorl has two triradii. Loops are further subclassified as radial or ulnar loops. This depends upon whether the loop opens on the radial or ulnar side of the finger.

The *ridge count* expresses the size of finger pattern. It is the number of ridges between triradius to the core of the pattern. The count of an arch is "zero" because it has no triradius. The sum of ridge counts of 10 fingers is called total ridge count. It is inherited as a multifactorial trait.

Francis Galton

(*Used from:* Karl Pearson's "The Life, Letters, and Labors of Francis Galton".)

Palmar Patterns

Herein, one has to locate four digital triradii, placed at the distal border of the palm and the axial triradius. The axial triradius usually lies near the proximal side of the palm but may be displaced distally in some disorders like Down syndrome or in trisomy 13. Between digital triradii, interdigital patterns may be seen in the form of recurving ridges. Thenar and hypothenar patterns may also be present (Fig. 12.4A). The axial triradial displacement indicates a possibility of an abnormal condition. The displacement of axial triradius is often expressed, either as a fraction of total palm length or as atd angle (Fig. 12.4B). Flexion creases, referred to as heart, head and life lines in palmistry form during the same period as dermal ridges. About half the Down syndrome patients

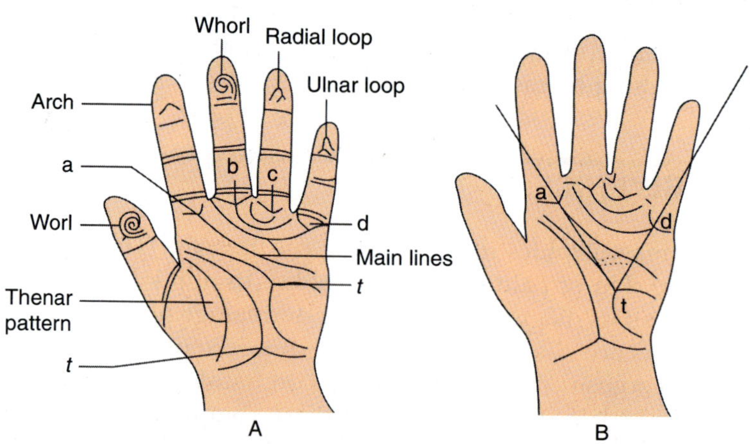

Figure 12.4 (A) The dermatoglyphic pattern of hand. Axial triradius *t* (*t'* its distal location). Digital triradii are a, b, c and d. (B) Measuring "atd" angle. If there is more than one axial triradius, the distal triradius is used.

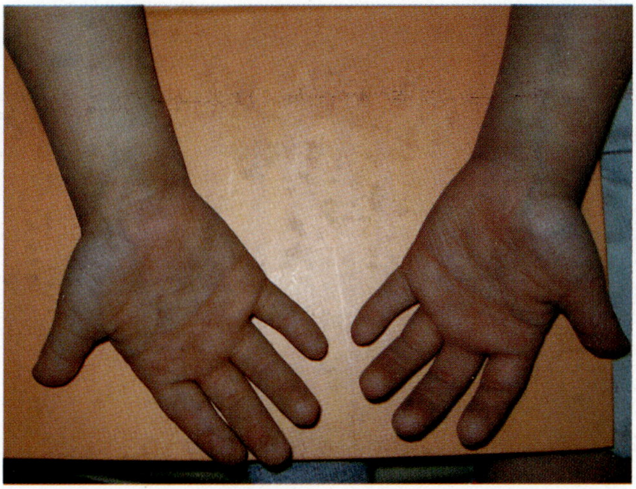

Figure 12.5 Simian crease seen on both palms in a Down syndrome case.

show a unique feature, i.e. single transverse crease on palm, often called *simian crease* (Fig. 12.5). However, this may be seen in about 1% of the normal persons. A variant type of pattern called *Sydney line* is seen in about 10% normal individuals. Sydney line represents a proximal crease of the palm, but distal crease is also present.

Plantar Patterns

There are sole patterns. They are less extensively studied as compared to palm patterns. Among them, tibial arch pattern in hallucal area is observed in 50% of Down syndrome patients.

Applications of Dermatoglyphics

1. In genetics, it is used as a supportive investigation for completing a diagnosis.
2. It helps in the determination of twin zygosity.
3. In criminology, it is used to identify the suspected criminal.

Summary

Population Genetics deals with the study of genes, distribution of genes and genotype in population.

Genetic Epidemiology: It deals with the distribution, aetiology and course of some heritable diseases in population.

Hardy–Weinberg's Law: It states that "gene frequencies" in a population remain constant from generation to generation if no evolutionary factors (e.g. migration, mutation, selection and drift) are operating.

Continued

Summary—cont'd

Beanbag Genetics: Beans of two colours, black and white, are present in a bag (total beans form a gene pool). If you draw two beans each time, one can expect two black, one black and one white or two white beans. Considering p and q as their frequencies p^2 (two black) + 2pq (one black, one white) + q^2 (two white). This in simple words has the algebraic formula; $p^2 + 2pq + q^2$.
Calculating Gene Frequency: Gene frequencies are expressed as fractions of unity, i.e. 1, the same can be derived with the algebraic formula $p^2 + 2pq + q^2$.
Factors influencing Hardy–Weinberg's equilibrium are:

1. *Mutation:* The mutation rate can be calculated by (*i*) direct or (*ii*) indirect method
2. *Mating pattern:* Random or non-random mating
3. *Genetic drift*
4. *Migration*

Eugenics: To establish best characters in population.
Twins: Monozygotic twins/identical twins and dizygotic twins
Abnormal Twining: Conjoined twins.
Chimaera: An individual having cells derived from two different zygotes, e.g. dizygotic twins with exchange of blood cells, dispermic chimaera.
Cloning: Dolly—the first cloned sheep, in 1997.
Possible use of cloning and stem cells is in heart diseases, diabetes, hypertension, spinal injuries, etc.
Dermatoglyphics: This means writing on skin. It involves study of ridge patterns on palms, soles and digits.
Palmar patterns: Four digital triradii and axial triradius are located, an "atd" angle is measured. Simian crease, a transverse crease on palm, is seen in Down syndrome patients. Sydney line represents a proximal crease of palm.
Plantar patterns: A tibial arch pattern in hallucal area is seen in 50% Down syndrome patients.
Applications: Dermatoglyphics can help in diagnosis, determination of twin zygosity and in criminology.

QUESTION YOURSELF*

1. What is population genetics?
2. What is Hardy–Weinberg's law?
3. What is "eugenics"?
4. What is "dermatoglyphics"?
5. What is simian crease?

*See page 266 for Answers.

Prenatal Diagnosis

13

COMPETENCY AN 81.1 Describe the various methods of prenatal diagnosis.
COMPETENCY AN 81.2 Describe the indications, process and disadvantages of amniocentesis.
COMPETENCY AN 81.3 Describe the indications, process and disadvantages of chorionic villus biopsy.

LEARNING OBJECTIVES

At the end of this chapter the students should be able to understand:

- PCPNDT Act
- Importance and indications for prenatal diagnosis
- Various procedures available for prenatal diagnosis
- Merits and demerits of these prenatal diagnostic procedures (techniques)

KEY WORDS

PCPNDT Act, Amniocentesis, CVS, Foetal blood sampling, Preimplantation diagnosis

Once we realise that genetic disorders hardly have any cure, we have to quickly think of preventive measures. With advances in diagnostic techniques in human genetics, it is now possible to diagnose many of the genetic disorders in utero. *Prenatal diagnosis* forms an integral step in genetic counselling. In fact, for couples at risk of a disorder, it is desirable to consider, plan and discuss prenatal diagnosis even before pregnancy. Discussion and planning beforehand will eliminate hurried procedures and emotional trauma as well.

Let us now consider the following situations that warrant prenatal diagnosis:

1. It is essential for a genetic disorder in which treatment is either absent or unsatisfactory.
2. Disorder in which an accurate prenatal diagnostic test is possible.
3. Risk to the pregnancy is sufficiently high.
4. The genetic disorder itself is severe enough to warrant termination of pregnancy.

5. Lastly the termination of pregnancy should be acceptable to the concerned couple.

In the following cases, prenatal diagnosis is a must:

1. Maternal age above 35/40 years.
2. If one of the parents is a balanced translocation carrier.
3. In case of an autosomal or X-linked recessive metabolic disorder that is severe but detectable prenatally.
4. Couple already has one child with a neural tube defect.

Having considered the indications for prenatal diagnosis, let us see how we should go about it. In other words, let us consider approaches to prenatal diagnosis. Of the various procedures available, two widely used ones are described in greater detail, the rest have been dealt with briefly:

1. Amniocentesis
2. Chorion villous biopsy
3. Ultrasonography
4. Foetoscopy
5. Foetal blood sampling
6. Maternal blood screening
7. Preimplantation diagnosis

AMNIOCENTESIS AN 81.1, 81.2

It is one of the prenatal diagnostic procedures with wide applications. Indications for the procedure are the same as those for prenatal diagnosis mentioned earlier.

Time

The ideal time to undertake this investigation is between 14 and 16 weeks when a sufficient amount of amniotic fluid is available for tapping, without harming the conceptus. This also ensures relatively easier acceptance of termination of pregnancy with an unfavourable outcome of amniocentesis results, around 18 weeks or so. Beyond this time, the patient's attitude towards termination of pregnancy alters because the foetal movement starts.

Procedure

Under ultrasound control, placental localisation is done (Fig. 13.1). Then under local anaesthesia, the fluid is tapped per abdomen avoiding injury to the placenta. A clear tap, not a blood-stained one, must be ensured. About 10–20 cc of fluid is taken out and is subjected to analysis in the laboratory. The cells and fluid are separated by centrifugation. The cells can be studied directly or subjected to culture studies for obtaining foetal karyotype.

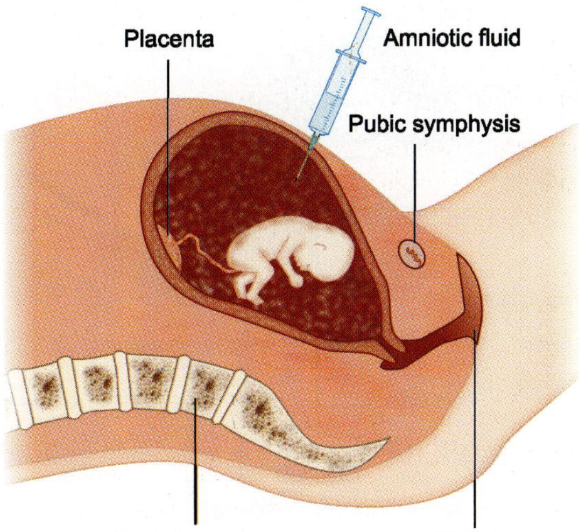

Figure 13.1 Amniocentesis, a prenatal diagnostic technique used in second trimester of pregnancy.

The fluid component is subjected to biochemical analysis for estimation of various ingredients such as α-foetoproteins (AFP). The results of the culture study take about 2–3 weeks or may be more. The patient and/or her relatives should be made aware of this fact.

Risk to the Pregnancy

The procedure carries 1%–1.5% risk of abortion, which is not significantly high. However, a repeat amniocentesis increases the risk to about 9%.

Perinatal Problems

The neonatal respiratory distress is increased almost three-folds. There appears to be an increase in postural orthopaedic problems such as talipes equinovarus (TEV) or congenital dislocation of hip. This procedure may slightly increase the risk of rhesus sensitisation, perinatal mortality and even antepartum haemorrhage.

CHORION VILLOUS BIOPSY

AN 81.1, 81.3

In this procedure, chorionic villi are aspirated with the help of canula, which is introduced through the cervix uteri (Fig. 13.2). The procedure is done under ultrasound control. The ideal time to perform chorion villous sampling (CVS) is 8–10 weeks period. However, it may be undertaken till almost 12 weeks.

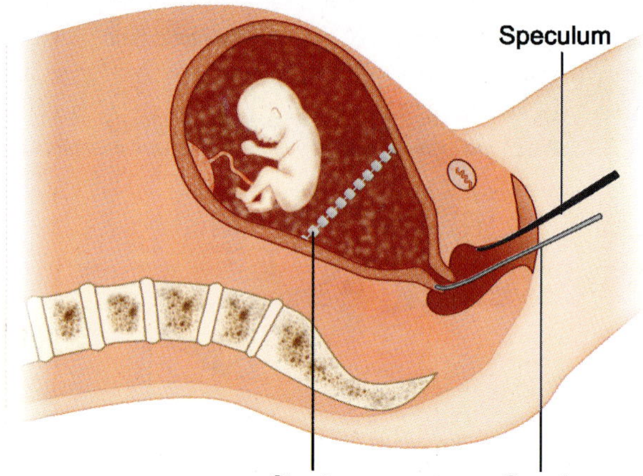

Figure 13.2 Chorion villous sampling for prenatal diagnosis in the first trimester of pregnancy.

Merits

1. As compared to amniocentesis, CVS claims an advantageous position because it is possible at a much earlier stage of gestation and is easily accepted by patients.
2. Faster result is possible because chorion villi contain enough cells under mitosis so as to permit chromosome analysis without culture.
3. If the results indicate abnormality in CVS, then termination of pregnancy is safer and simpler in first trimester than after amniocentesis (around 18 weeks), which amounts to second trimester abortion.

CVS is being accepted widely now. However, it carries a greater risk of abortion than amniocentesis. With experienced hands, it is still a safer procedure.

Problems of Foetal Chromosome Analysis

1. The first attempt towards culture of amniotic fluid or CVS may fail and investigation may be required to be repeated. Another alternative could be obtaining foetal blood sample by foetoscopy and culturing it.
2. From a cytogenetic point of view, maternal cell contamination may mislead you to diagnose a male foetus as a female one or mosaic with 46, XY/46, XX cell lines.
3. Mosaicism observed in foetal karyotype is little difficult to interpret as to whether it is true or pseudomosaicism.

Pseudomosaicism may have arisen with abnormal cell line appearing as (*i*) an artefact during culture and (*ii*) a result of maternal cell contamination. Usually, the sample is divided and multiple cultures are set with a single sample. If all the culture vessels show mosaicism, then it goes in favour of true mosaicism. However, if only one vessel shows mosaicism, in all probabilities it is likely to be pseudomosaicism. Another possibility being that the abnormal cell line may be from an extra embryonic tissue.

In short, while interpreting foetal chromosome analysis, all these factors have to be considered before putting the diagnosis. In similar manner, the parents should also be made aware of these difficulties in interpretation of foetal karyotype.

ULTRASONOGRAPHY

AN 81.1

The underlying principle in this procedure is that the echoes generated by the reflection of ultrasound waves are displayed in one of the two ways:

1. **B (brightness) Mode:** In this, a cross-section of the anatomy is created as transducer is moved across an area.
2. **Real Time Imaging:** In this, repetitive B-mode images are generated in rapid sequence, allowing appreciation of motion.

Basically, ultrasound serves as an ancillary to amniocentesis. It is helpful in the following ways:

1. Localisation of placenta in amniocentesis or CVS
2. To ascertain gestational age
3. Exclude multiple pregnancy
4. To recognise defects like anencephaly (Fig. 13.3), spina bifida, microcephaly, hydrocephalous, etc.
5. Severe limb defects are also evident on ultrasound

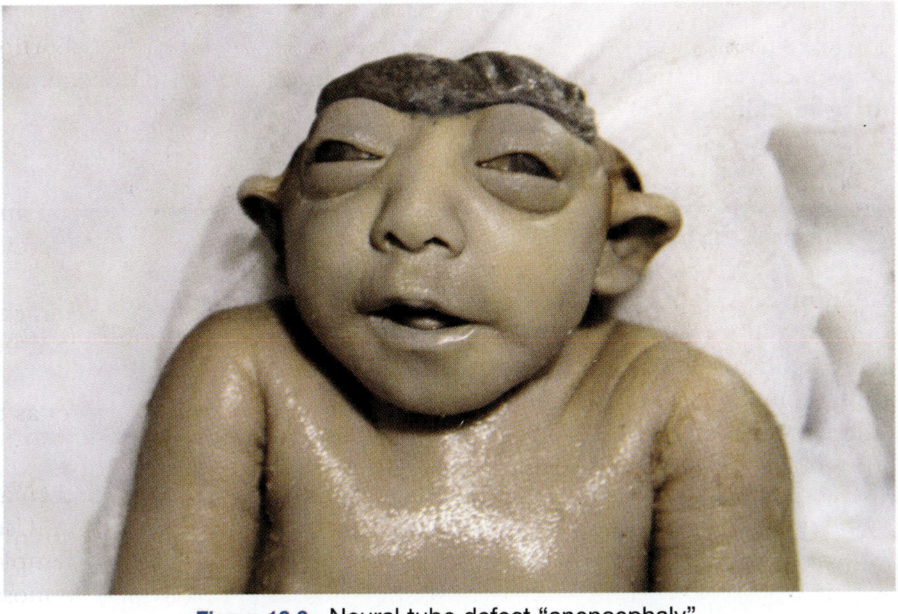

Figure 13.3 Neural tube defect "anencephaly".

FOETOSCOPY

AN 81.1

The procedure involves visualisation of foetus using a fibre optic self-illuminated instrument called foetoscope. It is inserted in the amniotic cavity under local anaesthesia. It is usually done around 18–22 weeks of gestation. With this, one can detect limb malformations, facial defects (cleft lip, cleft palate, ear defects) or defects involving the genitals.

The procedure carries a risk of abortion to the tune of 3%–5%. Foetoscopy is useful in obtaining foetoscopic skin biopsy and foetal blood sampling.

FOETAL BLOOD SAMPLING (FBS)

AN 81.1

It can be done in two ways:

1. Placental aspiration (indirect tap)
2. Sampling under direct vision

In the former technique, both maternal and foetal blood cells are mixed and need to be separated before sample processing. In the second case, sample is obtained under direct vision using a foetoscope. Both techniques carry about 10% risk of abortion. There are number of conditions in which FBS is needed to make prenatal diagnosis. They are as follows:

1. Sickle cell disease
2. Thalassaemias
3. Haemophilia A
4. Duchenne muscular dystrophy
5. Immune deficiency disorders

However, considering the high risk of abortion, one should use the investigation more judiciously.

MATERNAL SERUM SAMPLE

AN 81.1

Estimation of AFP in maternal serum is used as a screening test for the detection of neural tube defect (Fig. 13.3). This test is advocated for all pregnant women, realising the fact that about 90% babies with a neural tube defect are born to couples having no family history of such disorder.

Maternal serum shows AFP increment during 16–18 weeks of gestation. Elevated AFP in maternal serum is encountered in other conditions, e.g. twin pregnancy and missed or threatened abortion. Having noted elevated AFP, the patient is referred for ultrasonography and subsequently amniocentesis.

PREIMPLANTATION DIAGNOSIS AN 81.1

It is a technique that is still being developed, and it offers a reproductive option to a couple. It involves egg retrieval from the female followed by in vitro fertilisation (IVF). The fertilised oocyte is allowed to develop in vitro up to 8 cell stage. A single cell (blastomere) from this group is removed, its DNA extracted and amplified by PCR and then analysed to see if there is genetic disorder. If the analysis does not reveal any defect, the conceptus is implanted into the mother's womb. In X-linked recessive traits such as Duchenne muscular dystrophy, the preimplantation diagnosis is used to determine sex of conceptus (since only males are affected).

Demerits and Limitations

1. Despite PCR even in the best hands, procedure using single cell meets a failure rate of 10%–20%.
2. There is a significant risk of false results because of contamination. Hence, it is safe that an adverse result of preimplantation diagnosis should be followed by invasive prenatal diagnosis using CVS for confirmation.

RECOMBINANT DNA

With advances in diagnostic techniques, it is now possible to use recombinant DNA technology in prenatal diagnosis. DNA probes are available for prenatal diagnosis of haemoglobinopathies. May be in future recombinant DNA technology will replace other prenatal diagnostic procedures.

In conclusion, one can say that these techniques may be put together under a prenatal diagnostic programme. Such programme can be organised forming a team of geneticist, laboratory scientist and obstetrician with proper communication among them. Such an organised activity would serve to satisfy the needs of society. Varied opinions are expressed on the issue of termination of pregnancy with legal as well as social implications. They need to be assessed fully before one advises for MTP.

Summary

Prenatal Diagnosis forms an integral step in genetic counselling. In Maharashtra, it is governed by PCPNDT Act to prevent malpractices.

Indications:

1. The genetic disorder that has no satisfactory treatment.
2. Disorder that can be diagnosed accurately in prenatal stage.
3. Pregnancy at high risk (considering pedigree).
4. Severe disorder warranting termination of pregnancy.

Continued

Summary—cont'd

Prenatal diagnosis is MUST in:

1. Maternal age above 35 years.
2. One of the parent is translocation carrier.
3. Couple already has a child with disorder.

Prenatal diagnostic tests are:

1. ***Amniocentesis:*** Time 14–16 weeks of gestation. 10–20 cc clear amber-coloured amniotic fluid is withdrawn with aseptic precautions under ultrasound control. The fluid is centrifuged to separate cells, which are cultured to study chromosomes, and is subjected to biochemical analysis. Risk to pregnancy is 1%–5%. Perinatal problems (e.g. respiratory distress or TEV, CDH, Rh sensitisation) may follow.
2. ***Chorion villous biopsy:*** Few chorionic villi are aspirated under vacuum pressure through canula passed up the cervix with ultrasound control. The villi are washed and the cells are subjected to karyotyping. It is done around 8–10 weeks. So termination of pregnancy, if required, is relatively safe and easy.
3. ***Ultrasonography:*** It is almost routine in obstetric practice. It helps to monitor growth of the foetus, detection of anomalies, if any, to localise placenta in CVS or amniocentesis.
4. ***Foetoscopy:*** Using foetoscope one can visualise foetus. Done around 18–22 weeks. Detection of malformations, if any, can be done.
5. ***Foetal blood sampling:*** By placental aspiration or by direct method, foetal blood sample is drawn and karyotyping as well as DNA analysis of it is done for detection of diseases like DMD, sickle cell disease, haemophilia, etc.
6. ***Maternal serum sample:*** Estimation of AFP helps detection of neural tube defects. It is done around 16–18 weeks.
7. ***Preimplantation diagnosis:*** Involves IVF; at 8 cell stage, a single cell (blastomere) is removed. Its DNA is extracted and analysed for detection of genetic disorder.
8. ***Recombinant DNA:*** DNA probes for prenatal diagnosis of haemoglobinopathies are now available.

QUESTION YOURSELF*

1. When is the prenatal diagnosis warranted?
2. What is amniocentesis?
3. When is amniocentesis done and why particularly during that period?
4. What is the risk to the pregnancy after amniocentesis?
5. What are possible perinatal problems after amniocentesis?
6. What are the advantages of CVS over amniocentesis?
7. What are common problems of foetal chromosome analysis?

*See pages 266–267 for Answers.

Genetic Counselling

14

COMPETENCY AN 75.5 Describe the principles of genetic counselling.

LEARNING OBJECTIVES

At the end of this chapter the students should be able to understand

- Define genetic counselling
- Understand and follow the steps in the process of genetic counselling
- Calculate recurrence risk of the genetic disorder

KEY WORDS

Laws of addition and multiplication, Binomial distribution, Bayes' theorem, Screening programmes

AN 75.5

Genetic counselling is defined as a process in which patients or their relatives at the risk of a genetic disorder are made aware of the consequences of the disorder, its transmission and the ways by which this can be prevented or mitigated. Genetic counselling includes the following steps:

1. "Accurate" diagnosis of disorder.
2. Advising the patient and/or the family members with survey of relatives with similar complaints for disease confirmation or in otherwise normal relatives for carrier detection.
3. Management of the disorder, either curative (if possible) or supportive.

To complete accurate diagnosis, the following procedure should be followed:

1. History
2. Pedigree analysis
3. Estimation of risk
4. Transmitting information
5. Management

HISTORY

AN 75.5

A proper record of the history of patients is necessary:

1. This includes both present and relevant past history.
2. Family history includes sibs and other relatives also. Kindly note, if there is any other person in the family with a similar problem.
3. Obstetric history of the mother includes exposure to teratogens (drugs, X-rays) in pregnancy. History of abortion or stillbirth, if any, should be recorded.
4. Enquiry should be made about consanguinity, as it increases the risk especially in autosomal recessive disorders.

PEDIGREE CHARTING

AN 75.5

At a glance, this offers the state of a disorder in a family in concise manner. Constructing a pedigree with proper interrogation, though time-consuming, is ultimately rewarding. It forms an indispensable step towards counselling. Table 6.1 shows symbols used in pedigree charting.

ESTIMATION OF RISK

AN 75.5

It forms one of the most important aspects of genetic counselling. It is often called recurrence risk. To estimate it, one requires taking into account the following points:

1. Mode of inheritance
2. Analysis of pedigree/family tree
3. Results of various tests such as linkage studies

In order to estimate risk, one has to work out the probability. The probability of an outcome is defined as the number or more precisely the proportion of times it occurs in a large series of events. Routinely, the probability is indicated as a proportion/fraction of one. Probability 0.25 or ¼ indicates that on average, the event will be observed on 1 in 4 or 25% of occasions.

Theory of Probability, and Laws of Addition and Multiplication

Law of addition

While considering the probability of two different events, it is essential to know whether they are mutually exclusive or independent. If the events are mutually exclusive, then the probability that either one or the other will occur equals the sum of their individual probabilities. This is called the law of addition.

Law of multiplication

It states that if the two events are independent, then the probability that both the first and second event will occur equals the product of their individual probabilities.

For example, consider outcome of the first pregnancy. The probability that the baby will be either a boy or a girl equals 1, i.e. $\frac{1}{2}+\frac{1}{2}$. On ultrasonography, it is revealed that the mother is carrying dizygotic twins. Now the probability that both the first and second twin will be boys equals $\frac{1}{2}+\frac{1}{2}=\frac{1}{4}$.

Binomial Distribution

The concept of binomial distribution was used in genetics for the first time by Jacob Bernoulli, 17th century mathematician, though originally the binomial theory was discovered by Sir Isaac Newton in 1676.

Assume that the problem to be solved is the distribution of boys and girls in two successive births. If the probability of a boy is $\frac{1}{2}$ and the probability of a girl is $\frac{1}{2}$, then the distribution of families with two boys, one boy or no boys in all the two child families is given by the expansion of $\left(\frac{1}{2}+\frac{1}{2}\right)^2$. Instead of $\frac{1}{2}$ if we assume the frequency one event (having boy) is p; the other probability $q = 1 - p$. The expansion of the binomial $(p + q)^2$ gives different possible combinations, i.e.

For two child families,

$$(p+q)^2 = p^2 + 2pq + q^2 \quad \text{and } p = q = \frac{1}{2}$$

$$p^2 = \text{families with 2 boys} = \left(\frac{1}{2}\right)^2 = \frac{1}{4}$$

$$2pq = \text{family with 1 boy and 1 girl} = \frac{1}{2}$$

$$q^2 = \text{family with 2 girls} = \left(\frac{1}{2}\right)^2 = \frac{1}{4}$$

Likewise, for the families with three children:

$$(p+q)^3 = p^3 + 3p^2q + 3pq^2 + q^3$$

$$p^3 = \text{families with 3 boys} = \left(\frac{1}{2}\right)^3 = \frac{1}{8}$$

$$q^3 = \text{family with 3 girls} = \left(\frac{1}{2}\right)^3 = \frac{1}{8}$$

$$3p^2q = \text{family with 2 boys and 1 girl} = 3\left(\frac{1}{2}\right)^2\left(\frac{1}{2}\right) = \frac{3}{8}$$

$$3pq^2 = \text{family with 1 boy and 2 girls} = 3\left(\frac{1}{2}\right)\left(\frac{1}{2}\right)^2 = \frac{3}{8}$$

The advantage of binomial expansion is that it includes all the possible combinations of the two events. For example, in a family of three children there are in all eight probabilities of sequences as under:

B = Boy	**G = Girl**
BBB	GBB
BBG	GBG
BGB	GGB
BGG	GGG

The probability of each sequence works out to be 1/8. The family with two sons and one daughter, the daughter might be born first, second or last. Meaning there are three possible sequences that account for total 3/8 probability of having two sons and a daughter.

Bayes' Theorem/Bayesian Analysis

First published in 1763, Bayes' theorem offers a method of assessment of the relative probability of each of the two alternatives. Let us consider, how it works in estimation of risk in an X-linked lethal disease, e.g. Duchenne muscular dystrophy (Fig. 14.1).

In this pedigree II_3 is the daughter of an obligate carrier of DMD gene. The prior probability that she is a carrier is $\frac{1}{2}$ and prior probability that she is not a carrier is $\frac{1}{2}$. She/II_3 has three normal sons. If she is a carrier, the conditional probability that all three sons would be normal is $\frac{1}{2}\times\frac{1}{2}\times\frac{1}{2}=\frac{1}{8}$; if she is not a carrier, the conditional probability that all the three sons would be normal is 1 (very close to 1, because she may have a new mutant son) (Table 14.1).

Now let us consider the joint probability, which is the product of the prior and the conditional probabilities. The joint probability that she is a carrier is $\frac{1}{2}\times\frac{1}{8}=\frac{1}{16}$. The joint probability that she is not a carrier is $\frac{1}{2}\times 1=\frac{1}{2}$.

The posterior probability that she is a carrier is

$$\frac{1/16}{(1/16)+\left(\frac{1}{2}\right)}=\frac{1}{9};\ \frac{\text{Joint probability that she is a carrier}}{\text{Joint probability as carrier + Joint probability as non-carrier}}$$

$$= \text{Posterior probability}$$

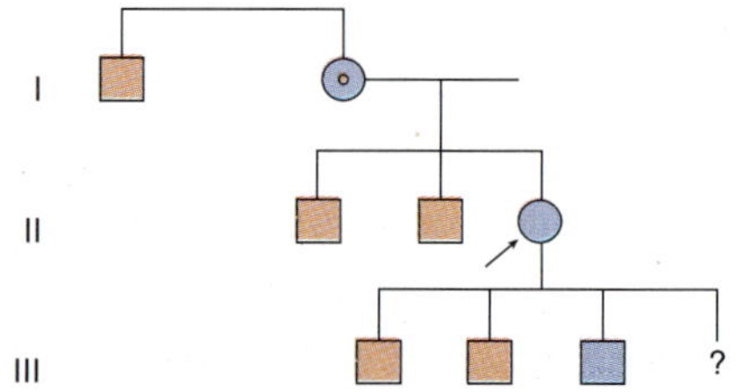

Figure 14.1 Pedigree of an X-linked recessive trait.

Table 14.1 Bayesian Calculations from Fig. 14.1

	II_3 Is a Carrier	II_3 Is Not a Carrier
Prior probability	1/2	1/2
Conditional probability	1/8	1
Joint probability	1/16	1/2
Posterior probability	1/9	8/9

The posterior probability that she is not a carrier is 8/9.

Now applying the posterior probability that II_3 is a carrier, the risk that her next child will be affected male is $1/9 \times 1/4 = 1/36$. This is remarkably below the prior probability of 1/8 if we do not consider/know about her children.

Autosomal recessive trait

Let us consider autosomal recessive trait, cystic fibrosis. Its pedigree is shown in Fig. 14.2.

As per Fig. 14.2, II_2 and II_3 each have an affected sib. This means:

Probability II_2 is a carrier = 2/3, probability II_3 is a carrier = 2/3. Probability both are carriers = $2/3 \times 2/3 = 4/9$. Therefore, the probability that next child will be affected is $1/4 \times 1/4 = 1/16$; much below the originally expected risk of $1/4 \times 4/9 = 1/9$ (Table 14.2).

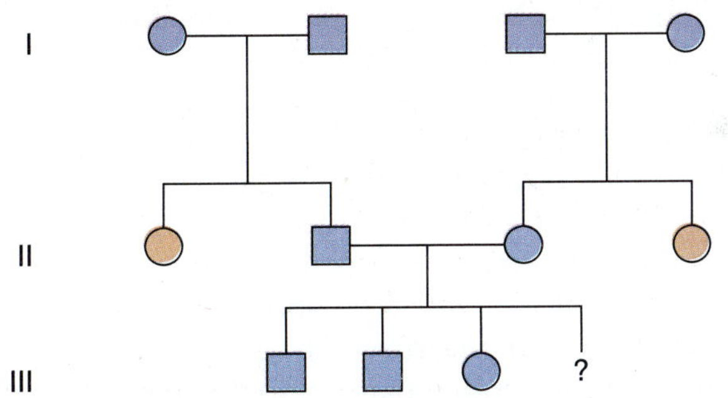

Figure 14.2 Pedigree of an autosomal recessive trait.

Table 14.2 Bayesian Calculation from Fig. 14.2

Probability	**Both Are Carriers**	**Both Are Not Carriers**
Prior probability	4/9	5/9
Conditional probability with three normal children	$(3/4)^3$	1
Joint probability	3/16 = 0.19	5/9 = 0.56
Relative probability	0.19/0.75 = 0.25	0.56/0.75 = 0.75

TRANSMITTING INFORMATION AN 75.5

After completing the diagnosis, pedigree charting and estimation of risk, the next most important step is of communicating this information to the consultants. This important function involves various factors. These factors are often not taken seriously, but are of crucial importance in genetic counselling. These are as follows:

1. Psychology of the patient;
2. Emotional stress under prevailing circumstances;
3. Attitude of family members towards the patient;
4. Educational, social and financial background of the family members;
5. Gaining confidence of consultants in subsequent meetings during follow-up;
6. Ethical, moral and legal implications involved in the process;
7. Above all, communication skills to transmit facts in an effective manner, i.e. making them more acceptable and palatable.

Now, the role of a genetic counsellor is to render help to consultants enabling them to take decisions. Should this be a directive? It is difficult to comment on this issue. In strict sense, "counselling" cannot/should not lead to directive advice. Also difficulty is encountered in making the relatives aware of the probabilities that are often complicated in certain situations. These are not well-understood by the people counselled.

MANAGEMENT AN 75.5

In genetics, "treatment" implies a very limited scope. It naturally aims for prevention rather than cure. In fact for most of the genetic disorders, cure is unknown. Treatment is therefore directed towards minimising the damage by early detection and preventing further irreversible damage, for example, in PKU. This disorder is characterised by a deficiency of phenylalanine hydroxylase enzyme, which is necessary for the conversion of phenylalanine to tyrosine. PKU, if not detected early, may lead to mental retardation owing to the involvement of nervous system at a later stage. The ideal situation would be early detection of the disease followed by preventive measures, like giving the patient a diet free from phenylalanine and thus preventing damage to the nervous system.

In some other situations, the defective gene proves to be so in certain environment. This implies that the changes in environment shall mitigate gene expression. Here also, an ideal way would be to replace the defective gene by a normal one, but this is left for the future; maybe in years to come such replacements become a reality.

PREVENTIVE ASPECTS AN 75.5

In the present situation, the aim of a geneticist is chiefly to prevent genetic defect. This means that prenatal diagnosis of a disorder should be made and the pregnancy (with abnormal foetus) be terminated. Termination of pregnancy should, however, be acceptable to the couple seeking advice. With this background let us try to work out various possibilities in some of the genetic problems:

1. Problem of infertility or inability to get a normal child. The couple has two alternatives. Either they can think of adoption, in which case pre-adoption counselling is important. In such children, a careful clinical examination of the child is done to rule out the possibility of a genetic disorder, since the parental/family background of the child is unknown.
2. Another alternative for such couple would be to go in for an artificial insemination donor (AID). This is appropriate if the father has or is at the risk of an autosomal dominant trait. It is also advisable when both partners are carriers of an autosomal recessive disorder. However, AID is not indicated, if mother has an autosomal dominant or X-linked disorder.
3. Analysis of a given case may be achieved through genetic tests such as chromosome analysis or with the help of various biochemical carrier detection tests. The test results, if negative, shall reassure the consultants that they are not at risk of the disorder. However, in a given situation, after making prenatal diagnosis or by working out the probabilities one can offer the information to the parents. Ultimately, the decision regarding termination of pregnancy has to be made by the couple.

FOLLOW-UP IN GENETIC CLINICS AN 75.5

Although follow-up is essential in all the branches under the faculty of medicine, it is all more important for patients attending genetic clinics. So, it is desirable to arrange more follow-up interviews. This will make sure that consultants understand and remember the information passed on to them. In some families with genetic disorders, repeated follow-up visits to the genetic clinic become essential. These visits are aimed at preventing the disease in any other family member by a reproductive planning, prenatal diagnosis followed by termination of pregnancy, if necessary. For the family member with the genetic disorder, acceptance of the disease, treatment if possible and counselling towards a more palatable way to lead life may be suggested. For example, take a family with Down syndrome (21 trisomy). The couple should first accept this defect in their child. They should then be made aware of and referred to a school for mentally retarded children where the child can be trained properly. Simultaneously, the couple can be informed about the possibilities of prenatal detection of this disorder as well as carrier detection (translocation carrier) in parents. This will prevent another Down baby in family.

GENETIC SCREENING AN 75.5

It was unknown earlier, but now it forms a part of the public health programme. The aim of such screening programmes was to identify newborns with genetic disorders so that early detection and treatment of the disease could be undertaken.

SCREENING PROGRAMMES

AN 75.5

The criteria for these programmes will be as follows:

1. The disorder should be clearly defined.
2. It should have a reasonable frequency in the population concerned warranting screening.
3. Disorder should be preferably treatable.
4. The screening test should be less time consuming.
5. The test should be relatively inexpensive so that it can be applied on a large scale.
6. The test should be reliable, i.e. ideally it should have minimal false positive and no false negative results.

With these prerequisites, screening programmes can be organised for newborns or for pregnant women. In the latter, maternal serum can be screened for neural tube defects estimating alpha-foetoproteins. Higher values of alpha-foetoproteins signify a neural tube defect, while unusually low values indicate foetus with Down syndrome.

LEGAL IMPLICATIONS

AN 75.5

Several malpractice suits have emerged in recent years against physicians considered negligent. A short but true story regarding one such case would duly stress the emphasis on the reader's mind. It is regarding a genetic disorder called Tay–Sachs disease. In this condition, there is a deficiency of enzyme hexosaminidase A. The condition manifests in the form of idiocy, mental retardation and progressive blindness. This disorder is relatively frequent in Ashkenazi Jews. One such Jew couple got tested for Tay–Sachs disease. They were informed that both of them were non-carriers of Tay–Sachs disease. Subsequently, the couple gave birth to a baby. This baby had Tay–Sachs disease. The matter was taken to the court. The court declared this child's birth as "wrongful life".

In conclusion, one can say that genetic counselling is perhaps the most important part of genetics where genetic knowledge is applied to the improvement of human health. This is achieved through currently available sophisticated techniques of prenatal diagnosis and various screening programmes.

Ethical and Legal Issues in Genetics

A sizeable budgetary provision was made in Human Genome Project to deal with ethical and legal issues involved. The areas under focus being genetic testing, gene therapy, stem cell research and embryonic cells. The advances in prenatal diagnostic techniques, sonography, amniocentesis, chorion villous sampling, foetal blood sampling, preimplantation diagnosis has led to more precise prenatal diagnosis with option of termination of pregnancy with untoward outcome of the tests. In this ethical issue being woman's right to terminate pregnancy. The issue took a new dimension in 1990, i.e. person with disabilities may not be properly accepted by the society and hence prenatal diagnosis may lead to **selective termination** of pregnancy. In the same manner withdrawal of support to newborn with severe malformation, e.g. trisomy 13 or extensive neural tube defect. In this the main principal should be to help the parents to take guided decision giving the information, i.e. counselling (not a directive advice).

Ethical concerns have also been about genetic testing of carrier detection and presymptomatic testing. A genetic test performed on one individual (an index case) may reveal, risk information of his relative who may not wish to know about it. For example, testing an index case in autosomal dominant trait may indicate that one of the parents has transmitted the abnormal gene. The genetic risk is often wrongly perceived as unchangeable. This leads to unfair stigmatisation of individual and his family.

Involvement of non-genetic factors in causation of disease also should be explained to patient and relatives. As is true in medical practice all the medical information should be confidential respecting the privacy of concerned individual and family.

In recent past **preimplantation diagnosis** has been the focus of ethical debate, since this may also reveal sex of the conceptus. It is true that earlier it was used to prevent implantation of male embryo possessing X-linked recessive trait. Equally true is the fact that this may also lead to sex selection which is inappropriate. Surprisingly preimplantation diagnosis can also be used to select embryo carrying disease causing mutation. A case was reported in United Kingdom in which deaf parents deliberately conceived a deaf child by artificial insemination (an autosomal recessive trait). An embryo homozygous for mutation causing an **autosomal recessive deafness** was selected. This certainly puts the interest of parents and of the child in conflict. In yet another case preimplantation diagnosis done to select HLA (human leukocyte antigen) matched embryo that could later provide bone marrow cells for its older sibling suffering from **Fanconi anaemia.** In all these cases the individuals may feel their life being devalued.

Genetic testing in childhood has also been questioned. If the testing is diagnostic and intervention is possible for example, children at the risk of inheriting mutation causing **adenomatous polyposis coli gene** (APC gene). In these children (gene carriers) colonoscopy should begin by 12 years of age as it is life saving. A contrasting situation exists in Huntington disease. The detection through childhood screening does not have any preventive or therapeutic advantage rather it increases anxiety and stigmatisation. Therefore, the childhood genetic testing should be avoided unless there is clear clinical benefit through intervention is possible.

One of the most controversial and debatable ethical as well as legal issue has been "cloning" and "stem cell research". It is pertinent to note that reproductive cloning and therapeutic cloning are different, the former, i.e. reproductive cloning has been successful in mammals, however practicing it in humans is unanimously opposed by the entire scientific community. Equally true is the fact that therapeutic cloning has an application in "stem cell transplant" and can be useful in treating blindness and cancer.

Summary

Genetic Counselling: Process by which patients and their relatives are made aware of the consequences of a genetic disorder, its transmission and the ways to prevent/mitigate it.

Steps involved in genetic counselling

i) ***Accurate diagnosis:*** It involves (*a*) *history*—past, present, obstetrics history, family history; (*b*) *pedigree*; (*c*) *investigations*—pathological, hormonal and imaging.

ii) ***Survey of relatives:*** With similar condition, recurrence risk estimation.

iii) ***Management of the disease.***

Estimation of Risk: (*i*) Law of addition; (*ii*) law of multiplication; (*iii*) ***binominal distribution*** introduced by Jacob Bernoulli in 17th century, which is based on binominal theory discovered by Sir Isaac Newton in 1676; (*iv*) ***Bayes' theorem/Bayesian analysis***: It is a method of assessment of relative probability of each of the two alternatives. It takes into consideration the a) *prior probability*, b) *conditional probability*, c) *joint probability* and d) *posterior probability*.

Transmitting Information: Factors to be considered are (*i*) Psychology of the patient; (*ii*) Emotional stress; (*iii*) Attitude of family towards patient; (*iv*) Educational, social and financial background of the family members; (*v*) Gaining confidence of the patient and relatives; (*vi*) Ethical, moral and legal implications of the process; and (*vii*) Communication skills of the counsellor.

Management: Early detection and interaction is the key.

Preventive Aspects: Prenatal diagnosis and subsequent termination of pregnancy, if essential.

Follow-up: It is crucial in genetic clinics. This helps in preventing recurrence in the next generation through prenatal screening and reproductive planning.

Screening Programmes: Criteria to design programmes are:

1. Clearly defined disorder
2. Reasonable frequency in population
3. Disorder has treatment
4. Programme designed should be less time consuming
5. Cost effective and reliable

Legal aspects and ethical issues should be dealt carefully.

QUESTION YOURSELF*

1. What is genetic counselling?
2. What are the steps involved in genetic counselling?

*See page 267 for Answers.

15 Gene Therapy

LEARNING OBJECTIVES

At the end of this chapter the students should be able to understand

- What is gene therapy?
- How gene delivery is done with various methods
- Advantages and disadvantages of these systems.

KEY WORDS

Gene transfer techniques, Retroviral vectors, Candidate disease, Target tissues

Gene therapy involves replacement of a defective/abnormal gene into the cells of a patient who is deficient of the normal gene product. The technical basis of gene therapy is gene delivery, i.e. introducing the desired gene into the appropriate cells of the patient. The current wave of gene therapy research has gathered momentum because highly effective gene delivery systems have been developed, in particular those based on the use of retroviral vectors.

GOVERNING BODIES

In many countries, there are regulatory bodies to observe the technical, therapeutic and safety aspects of gene therapy. In the United States, National Institute of Health (NIH) has laid down guidelines for clinical trials of human gene therapy. Unless the protocol is approved by Food and Drug Administration (FDA) and Recombinant DNA Advisory Committee (RAC), it cannot be undertaken. In the United Kingdom, the Committee on the Ethics of Gene Therapy has recommended that all the gene therapy protocols must be approved by the hospital research ethical committees.

It has been universally accepted that "germ line gene therapy" should not be instituted. This implies that all the gene therapy protocols must focus only on the *"somatic cells gene therapy"*.

GENE DELIVERY

Gene therapy requires introduction of foreign DNA sequences with stable integration, gene expression and an appropriate regulation in the target tissue. The newly introduced gene can replace a missing gene. There are two strategies used to deliver genes—(*i*) ex vivo and (*ii*) in vivo transfer. In ex vivo transfer, cells are removed from the patient, an appropriate gene (DNA sequences) is introduced in these cells and then these genetically engineered cells are transplanted back into the patient's body. In in vivo approach, the desired gene is directly introduced into the target tissue.

GENE TRANSFER TECHNIQUES

Transfer of gene can be accomplished by following methods: (*i*) physical or (*ii*) biological (viral vectors).

Physical Transfection Methods

These include:

(a) Liposome-mediated DNA transfer
(b) Receptor-mediated endocytosis

Liposome-mediated DNA transfer

It involves complexing plasmid DNA (with foreign DNA) with liposomes and introducing it into the target cell (Fig. 15.1).

Advantages: Using this technique, one can introduce larger amount of DNA into the target cells than what is possible with viral vector systems. This can be as large as an artificially constructed minichromosome, which includes elements for regulation of gene expression apart from the particular structural gene. These elements regulate gene expression in physiologically controlled manner.

Disadvantage: The gene expression is transient; therefore, the treatment has to be repeated.

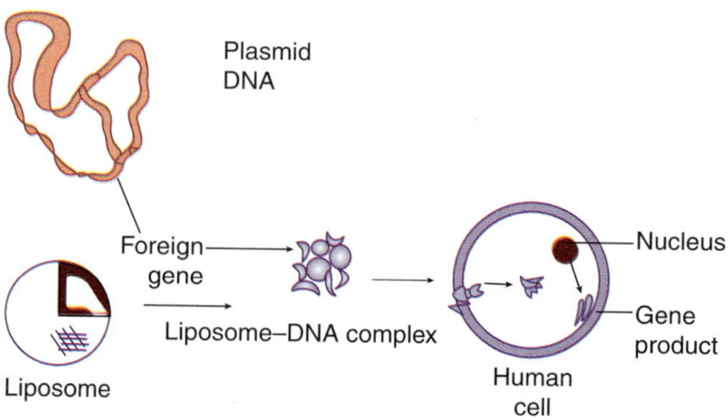

Figure 15.1 Liposome-mediated gene therapy.

Animation 5: Gene therapy. (*Available online at* www.medenact.com)

Receptor-mediated endocytosis

In this method, a complex is made between plasmid DNA containing foreign DNA and specific polypeptide ligands for which the cell has receptor. The DNA is targeted to these receptors. For example, DNA is complexed to a glycoprotein containing galactose. This will be recognised by the receptors on liver cells, which are specific to glycoproteins with a terminal galactose. This causes internalisation of the complex by endocytosis. The endocytic vesicles fuse with lysosomes, and here complex is degraded and foreign gene escapes from the lysosome, to be expressed. The rate at which it (foreign DNA) escapes from the lysosomes can be increased by inclusion of adenovirus or fusogenic influenza gene products.

Major advantage of the physical transfection method is that they are free from the risk of viruses. Liposome-mediated gene therapy is being actively tested on cystic fibrosis patients in UK and USA.

Viral Vectors (Biological Transfection Method)

In biological transfection methods, viral vectors form an efficient mode of gene delivery into the target cells (Fig. 15.2). Various viruses used include retroviruses, adenoviruses, herpes virus, adeno-associated virus, parvovirus, etc. The viruses are rendered replication deficient by removing encapsidation (ψ, psi) gene sequences.

Retroviruses

They are derived from the family of viruses that includes human immunodeficiency virus and oncogenic viruses capable of doing so in some species. The retroviruses are rendered incapable of replication by removing

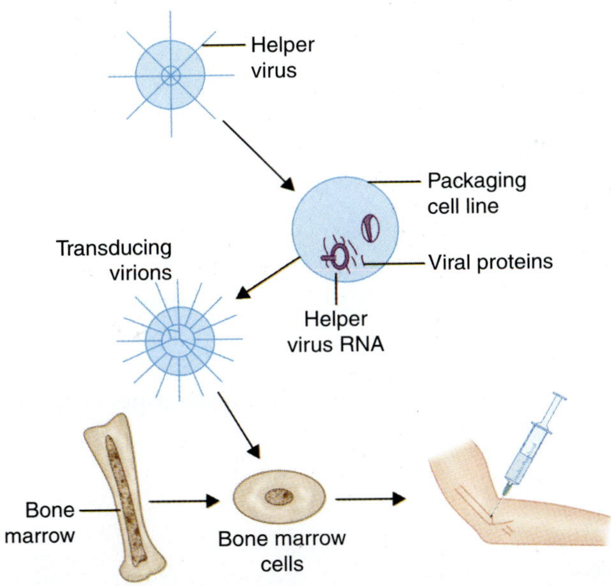

Figure 15.2 Retroviral gene therapy using bone marrow cells.

encapsidation sequences, which are essential for viral replication. Retroviruses integrate into the host DNA and make copy of their genome using reverse transcriptase enzyme. The provirus thus formed serves as a template for production of mRNAs for various viral gene products as well as the new genomic RNA of the virus. The retroviruses used in gene therapy need a couple of elements—(*i*) packaging cell line and (*ii*) helper virus.

1. **Packaging cell line:** It is the cell line that has been infected with the retrovirus that is genetically engineered to lack region of proviral DNA called packaging sequence.
2. **Helper virus/vector:** It consists of a retroviral provirus with its more than 90% of viral genomic material removed, leaving only minimal sequences essential to produce copies of the viral RNA along with the sequences necessary for packaging of the viral genomic RNA. This is a vector backbone in which foreign gene (DNA sequence) can be inserted. If this helper virus is introduced into the packaging cell line that contains provirus in which packaging sequences are missing, then RNA produced by the vector provirus can be packaged into viruses. These *virions* can be used to infect or more precisely called *transduce* the target cells.

Disadvantages: Demerits of retroviruses as vectors in gene therapy are as follows:

1. Only smaller DNA sequences (less than 7 kb usually) can be introduced, e.g. even if all introns were removed from the dystrophin gene to be used in gene therapy in case of Duchenne muscular dystrophy, still the gene would be large enough to be incorporated into the retroviral vector. To make it possible, large amount of sequences are deleted, still retaining relatively normal function. This is called mini dystrophin gene.
2. Retroviral vector can transduce only dividing cells and hence central nervous system disorders are not amenable to it.
3. The retroviruses can only be used in vitro.
4. Another demerit of retroviruses is that they are unstable.
5. They cannot be purified for use in gene therapy without reducing their capability to transduce target cells. This means contamination with the replication competent retroviruses is inevitable. They could serve as an oncogene, causing malignant transformation.
6. Controlling the levels of expression of the introduced gene is yet another difficult task.

Retroviral gene transfer can be effectively used in case of hepatocytes, haemopoietic stem cells (HSCs), fibroblasts myoblasts, endothelial cell, etc.

Adenoviruses

These are also used as vectors in gene therapy, especially in non-dividing cells, e.g. central nervous system cells.

Advantages:

1. These are stable.
2. Easily purified.

3. Suitable for targeted treatment of specific tissues, e.g. respiratory tract.
4. These can infect/transduce non-dividing cells.
5. These can carry larger DNA segments as big as 36 kb long.

Disadvantages:

1. They do not integrate into the host genome; therefore, expression of the introduced gene is unstable.
2. The gene expression is often transient.
3. By virtue of their infectivity, they can produce adverse effects secondary to infection.
4. Adenoviruses contain genes that can cause malignant transformation; hence, their use as vectors carries a menace of inducing malignancy.

Herpes virus

Herpes viruses are neurotropic viruses, which on suitable modification can be effectively used for gene therapy in central nervous system disorders.

Advantages: Herpes virus has a natural affinity for non-dividing cells and hence it is suitable for transfection of neurons. It can also be used in hepatocytes.

Disadvantages: These are as follows:

1. An immediate disadvantage of using herpes viruses as vectors is because of their toxic effects on the nerve cells and the following immune response.
2. Since herpes viruses do not integrate into the host genome, the expression of the introduced gene may be unstable.

Parvovirus and simian virus 40 are being considered for gene therapy in central nervous system and smooth muscle cells, respectively. Studies undertaken on adeno-associated virus (AAV) have revealed following *merits,* i.e. they (*i*) form a stable preparation, (*ii*) lack pathogenicity and (*iii*) have high efficiency of integration. *Disadvantages* of AAV as vectors include possibility of immune response.

Target Tissues

Insertion of a normal gene into the diseased tissue depends upon proliferative state of the tissue, an accessibility to gene manipulations and normal site of gene expression.

Liver: Hepatocytes are refractory to retrovirus in vivo; they are, however, more susceptible to transduction by retrovirus in vitro. Liver forms a suitable target organ owing to its rich vascularity. Hepatocytes are removed after partial hepatectomy. They are grown in culture, transduced with desired gene through retroviral vector and then returned via hepatic artery or portal venous system. However, there is risk of portal venous thrombosis that can lead to portal hypertension. This approach has been used for patients with familial hypercholesterolaemia caused by missense mutation of low density lipoprotein receptor gene (LDLR). This leads to reduction of LDL levels for a short term; the long-term benefit is still awaited. Other disorders involving liver in which similar approach could be considered are haemophilia A, α_1-antitrypsin deficiency and phenyl-ketonuria.

Muscle: Direct injection of foreign DNA into the muscle has met with reasonable success in terms of retention and expression of the foreign gene. As an alternative, myoblasts can be injected into the muscle. This results in their incorporation into the recipient muscle fascicles. This approach can be used in vitro to insert genes into the myoblasts that are totally unrelated to the muscle function. For example, factor VIII and human growth hormone.

CNS: Vector systems are being developed for CNS disorders. They consist of replication of defective neurotropic adenoviruses lacking El region. They are then made infective by growing them in the cells engineered to express El genes. Alternatively, one can transplant cells that have been genetically modified in vitro into specific regions of the brain, like caudate nucleus in Huntington disease.

Bone Marrow: Treatment of bone marrow disorders poses a problem because of the low frequency of stem cells. HSCs form an ideal target for gene therapy. However, currently many gene therapy efforts are aimed at more differentiated haemopoietic cells such as lymphocytes. Harvesting and inserting gene directly into the stem cells is possible using monoclonal antibodies that recognise cell surface marker CD_{34}. For ADA-deficiency SCID, lymphocytes are isolated from blood, grown under conditions promoting growth of T-lymphocytes (culture medium containing antibody to T-receptor and T-cell growth factor IL-2). ADA gene is introduced in T-cells using retroviral vector. Other target tissues include fibroblasts, endothelial cells, airways epithelial cells, glial cells, etc.

DISEASES AMENABLE TO GENE THERAPY

Diseases in which gene therapy can be conceived include both genetic and non-genetic diseases. They satisfy the following criteria:

Prerequisites of Candidate Disease

1. The candidate disease should be genetically recessive.
2. The defective gene is identified and cloned.
3. Precise regulation of the gene product is not required.
4. The cells bearing corrected/desired gene should have selective advantage over uncorrected cells.

Some shortcomings in selecting candidate disease for gene therapy include following difficulties: (*i*) disorders involving complex gene regulation and (*ii*) diseases in relatively inaccessible tissues like CNS. The gene therapy in such situations (CNS) has to be instituted before irreversible damage occurs.

Cancer

Gene therapy for cancer involves the introduction of tumour suppressor gene or inactivating an oncogene or use of immune cells and so on. Potential strategies for gene therapy in the cancer treatment are as under:

1. **Tumour suppressor gene:** Inserting a wild-type tumour suppressor gene, e.g. p53 or the gene involved in Wilms tumour.
2. **Blocking oncogene:** Blocking expression of an oncogene, e.g. by introducing the gene that encodes anti-sense K-RAS message.
3. **Suicide gene:** Insertion of a sensitivity or suicide gene into the tumour, e.g. introducing gene that encodes thymidine kinase gene of herpes simplex virus (HSVTK).
4. **Promoting immunogenicity of tumour:** This is achieved by introducing genes that encode foreign antigens.
5. **Use of genes for cytokines:** Enhancing immune cells to increase anti-tumour activity by inducing genes that encode cytokines.
6. Protecting stem cells from the toxic effects of chemotherapy by introducing the gene that confers **MDR-1** (multiple drug resistance-1).

Peripheral Vascular Disease

In persons with peripheral vascular disease, the arterial segments or the vascular grafts could be resurfaced by endothelial cells or smooth muscle cells in which anti-clotting agent genes have been incorporated.

Coronary Artery Disease

In persons having family history of an early coronary artery disease, gene therapy could be used to introduce LDL receptor. A particularly important area under investigation is the prevention of reocclusion after angioplasty. One approach involves use of mutant forms of tissue type plasminogen activator (TPA) having thrombolytic effect. This can be delivered by adenovirus to the specific tissue and can quickly lyse a clot. The other alternative being modifying endothelial cells so that they can secrete TPA. These genetically engineered endothelial cells can be implanted in the graft to prevent clotting. Preventing smooth muscle proliferation through genes may also be rewarding because it is supposed to be the principle cause of reocclusion.

Acquired Immunodeficiency Syndrome (AIDS)

Gene therapy for AIDS includes transducing or transplanting totipotent haemopoietic stem cells (THSCs). Inhibiting viral replication forms an alternative approach.

The first human gene therapy was instituted on September 14, 1990. A 4-year-old girl suffering from adenosine-deaminase (ADA)-deficiency SCID was given transfusion of her own peripheral blood T-lymphocytes that had been transduced ex vivo using retroviral-mediated gene transfer with the normal human ADA gene. She received 11 transfusions over the next 2 years, resulting in her intracellular ADA concentration to rise from an undetectable state to about 20%–30% of the normal value. Her T-lymphocyte count was also elevated to almost normal value. She subsequently showed an average infection rate.

Second ADA-deficient SCID patient receiving gene therapy was a 9-year-old patient with his treatment initiated in January 1991. Similar laboratory findings and clinical improvement was observed in this patient as was found in the first ADA-deficiency case.

Familial hypercholesterolaemia, a trait resulting from a defect in LDL receptor gene, leads to ischaemic heart disease in these individuals. Here, hepatocytes are removed from the patient and transduced with the retroviral vector containing human LDL receptor gene. Studies indicate that this technique is feasible with transduction efficiency rate close to 30%. Gene therapy trial for haemophilia B began in December 1991, in Shanghai, China. It was retroviral-mediated gene transfer of gene for factor IX using skin fibroblasts.

In cystic fibrosis, gene therapy aims at delivering CFTR or α_1-antitrypsin gene directly into the epithelial cells lining airways. The α_1-antitrypsin gene can also be directed towards hepatocytes. As lung tissue proliferates very slowly, it is more suitable for adenoviral vectors. However, this corrects only the pulmonary complications of cystic fibrosis. The other method used to deliver CFTR gene is liposome-mediated gene transfer. At NIH, the first human gene therapy trial in cystic fibrosis was initiated on April 17, 1993.

FUTURE PROSPECTS

For gene therapy to become a widely used medical therapy, few hurdles need to be overcome. First, we need gene transfer vectors that can be injected directly into the patient. Second, the vector should integrate safely into a non-critical site on chromosomes or homologous recombination replacing defective gene. Finally, the introduced gene should have an ability to respond to physiological changes in blood or cellular metabolites. For example, in gene therapy for diabetes (insulin deficiency), rise or fall in blood glucose levels should be sensed and responded to by an appropriately engineered insulin gene. The future of gene therapy is promising, and it is likely that this technology will be applied to treat a wide variety of diseases in the years to follow.

Summary

Gene Therapy involves replacement of a defective/abnormal gene. Gene delivery is based on the use of retroviral vectors.

Governing bodies include NIH (National Institute of Health), FDA (Food and Drug Administration), RAC (Recombinant DNA Advisory Committee), Committee on Ethics in Gene Therapy, etc.

Gene delivery is done by either ex-vivo or in vivo strategy.

Gene Transfer Techniques include (*i*) ***physical transfection methods*** such as liposome-mediated endocytosis; (*ii*) ***retroviral transfection method***—this needs packaging cell line and helper virus. Adenoviruses are used in non-dividing cells, e.g. CNS.

Target Tissues: It depends upon proliferation state of tissues, accessibility to gene manipulation and site of gene expression. These include liver, muscles, CNS, bone marrow.

Diseases Amenable to Gene Therapy: (*i*) ***Cancer*** with the use of tumour-suppressor gene, blocking oncogene, suicide gene, promoting immunogenicity of tumour, use of genes for cytokines; (*ii*) ***peripheral vascular disease;*** (*iii*) ***coronary artery disease;*** (*iv*) ***AIDS***; (*v*) ***cystic fibrosis***, etc.

Future Prospects: Injections of gene transfer vectors directly, to develop safe and cost effective gene delivery that responds to physiological and metabolic changes in blood.

QUESTION YOURSELF*

1. What are physical transfection methods of gene delivery?
2. When was the first human gene therapy instituted?

*See page 267 for Answers.

16 Stem Cell Therapy

LEARNING OBJECTIVES

At the end of this chapter the students should be able to understand

- What are stem cells?
- Different sources to obtain stem cells
- Application of knowledge of stem cells in clinical practice

KEY WORDS

BMSCs (bone marrow stem cells), PBSCs (peripheral blood stem cells)

"Stem cells" and stem cell therapy have opened altogether a new chapter in healthcare. These cells (which are "pluripotent"), under suitable conditions, can differentiate into any type of cells. In other words, any organ can be formed through these cells. Thus, in future, we can think about replacement of a diseased organ with a fresh one. This will revolutionise the field of organ transplantation. Recently a group of researchers from Kolhapur, Western Maharashtra, India, have proposed that endometrium is a rich source of stem cells. If this holds true, we would not need to bank upon umbilical cord cells. Currently, there are institutions preserving umbilical cords, of course at sumptuous cost. This, i.e. uterine source, would be a much cost effective way to "stem cell therapy".

Stem cells have seemingly endless self-renewal potential. Their undifferentiated state makes it possible. From a single cell, many healthy cells can be formed to replace damaged cells of adult organism. This forms the basis of cell-based therapies as newer treatment modality. This has also been called "regenerative medicine" and is the most fascinating area of biology.

STEM CELL

It has an ability to divide infinitely; under appropriate conditions, it can give rise to many different cell types to form desired tissues.

The stem cells are of two types:

i) Embryonic stem cells
ii) Adult stem cells

Embryonic Stem Cells

They are derived from the inner cell mass/embryoblast of the blastocyst. They have potential to infinitely divide symmetrically without differentiating, i.e. long-term self-renewal. They are "clonogenic", i.e. each single cell can give rise to a colony of genetically identical clones, having same functional potential as the original cell (Fig. 16.1).

i) The embryonic stem cell lines are established from embryos shortly after fertilisation.
ii) To create an embryonic stem cell line, an embryo must be separated into individual cells.
iii) A single cell from the embryo is placed in a dish and provided with growth factors as well as nutrients. This stimulates it to divide.

Embryonic stem cell (ESC) can be grown into any tissue, e.g. heart muscle, liver, kidney, muscle, etc. However, there is lot of ethical debate on the use of ESCs for research and therapy.

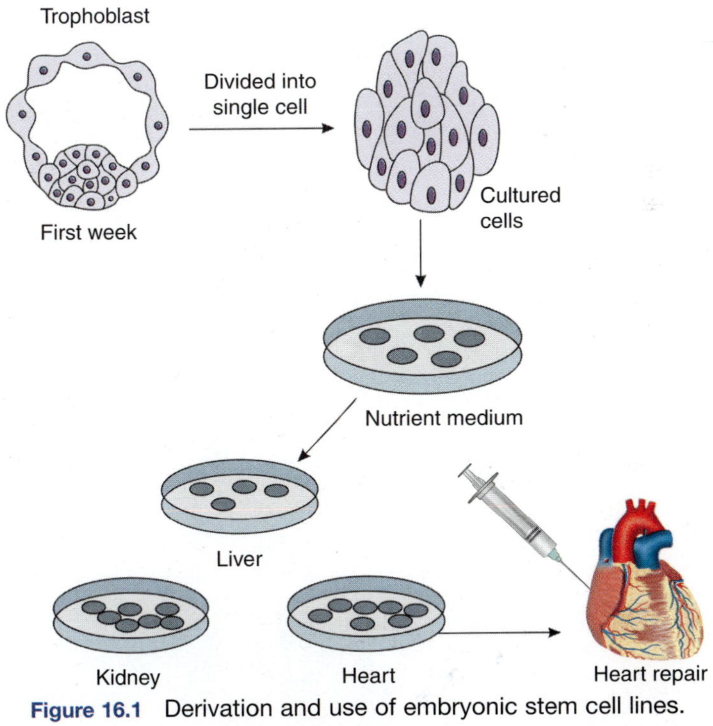

Figure 16.1 Derivation and use of embryonic stem cell lines.

Animation 6: Stem cell therapy. (*Available online at* www.medenact.com)

Potential uses of stem cell can be in many conditions, such as:

i) Stroke, traumatic brain injury, Alzheimer disease
ii) Myocardial infarction
iii) Wound healing
iv) Osteoarthritis
v) Diabetes
vi) Muscular dystrophy
vii) Multiple-site cancers

Adult Stem Cells

It refers to undifferentiated cells found among differentiated cells in a tissue or an organ in adults. They can renew themselves and can differentiate, yielding major specialised cell type of the tissue/organ. Major role assigned to them is to maintain and repair the tissue/organ in which they are found. However, they are rare in tissues/organs in which they exist. It is estimated that one in 10,000–15,000 cells in the bone marrow is haemopoietic stem cell (HSC).

The tissues reported till date to have stem cells are (*i*) bone marrow, (*ii*) peripheral blood, (*iii*) brain, (*iv*) spinal cord, (*v*) skeletal muscle, (*vi*) epithelia of digestive system, skin, cornea, retina, liver, pancreas, (*vii*) dental pulp. The list is growing everyday. In recent times, a great deal of excitement has been generated in research on adult stem cells. Their potential use in transplant has offered a ray of hope in many diseases (Fig. 16.2).

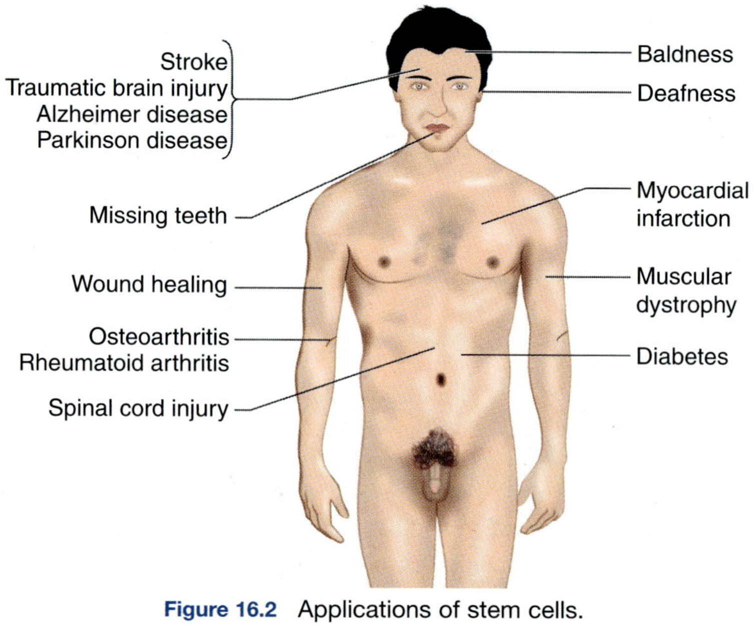

Figure 16.2 Applications of stem cells.

STEM CELL THERAPIES TODAY

Adult stem cell transplant:
Bone marrow stem cells (BMSCs)
Peripheral blood stem cells (PBSCs)
Umbilical cord blood stem cells

Bone Marrow Transplant

Perhaps the best-known stem cell therapy to date is the bone marrow transplant, which is used to treat leukaemia and other types of blood cancer as well as other blood disorders.

Peripheral Blood Stem Cell Transplant

Since peripheral blood stem cells (PBSCs) can be obtained from the blood drawn, it is easier to collect than bone marrow stem cells because bone marrow stem cells (BMSCs) have to be extracted from the bones. So, in short, PBSCs make a less invasive treatment option than BMSCs. However, PBSCs are sparse in the bloodstream and hence collecting enough of them to perform a transplant poses a challenge.

Umbilical Cord Blood Stem Cells Transplant

Recently, scientists and the healthcare professionals have focused their attention on multipotent stem-cell–rich blood found in the umbilical cord. It has proven useful in treating same type of health problems as those treated using BMSCs and PBSCs.

Cord blood banking

A few years ago, cord blood banking was the fancy of rich class. Considering the potential it holds, many entrepreneurs have stepped into this area. This has made it cost effective, and now it is being taken up by middle class also. Though stem cell therapy is still in nascent stage, it holds a promise in future.

Few institutes undertaking this in India include the following:

i) Life Cell India
ii) Reliance Life Sciences
iii) Cord Life Biotech

Indian Council of Medical Research (ICMR), an apex body in the field of medical education and healthcare, has formed guidelines for stem cell research in 2007. Now by 2012, the stem cell research has truly gathered momentum. It is evident from the fact that LV Prasad Eye Institute, Hyderabad has treated more than 800 patients using limbal stem cells for repairing corneal surface disorders. In civic/private-run hospitals in Mumbai, Dental

Stem Bank, Stem Cell Labs have been working since 2006. With global market of around $20 billion, India needs to enhance its current share.

SYNTHETIC CORNEA

We, the healthcare professionals, always promote "Eye donation"; still the number required and desired is much more as compared to people donating eye or pledging for it. Synthetic cornea could be a right solution to this problem. Synthetic cornea is prepared by taking out little collagen from the individual and then forming complete cornea out of it. In this way, it serves as an "autograft", and hence there is no fear of rejection.

ETHICAL, LEGAL AND SOCIAL ISSUES

Considering and addressing all these issues is not a simple task. There are many questions and arguments both in favour and against stem cell. Research, especially on ESCs, generates the following questions and arguments:

Questions

When does human life begin?
Is killing a zygote, killing a human being?
Frozen embryos in IVF are going to be discarded anyway; can they be used for research?

Arguments

Humans are created in the image of God before birth.
Human soul begins before birth.
So *NO* human zygote/embryo should be used for research.

RECENT PANDEMIC OF COVID-19

To treat the coronavirus patients, apart from antiviral drugs a couple of options have been tried.

1. **Plasma Therapy:** This involves using the antibodies in the plasma of a COVID-infected person who has recovered. His/her plasma still contains the antibodies which can combat the virus in another COVID-infected patient.
2. **PLX Therapeutics:** A biotech company Pluristem, based in Haifa, Israel, has developed PLX therapy. It involves Pluristem allogeneic placental expanded (PLX) cells. These cells suppress or reverse dangerous overactivation of the immune system which causes death in coronavirus patients. This was tried in patients at risk of death owing to respiratory failure and multiorgan failure. This triggered a ray of hope and in April 2020 this therapy was being tried at yet another place Holy Name Medical Centre in New Jersey, USA.

An intramuscular injection of these PLX cells is given. Once inside the body, these cells serve as a factory generating therapeutic proteins of two types:

i) One protein reduces the inflammation.
ii) Other modulates immune systems.

The cells are taken from a full-term placenta. These are then expanded to enormously large number with the developed technology. They claim that over 20,000 patients can be treated through cells derived from one placenta. A multicentric/multinational study is required to prove the efficacy of the PLX therapy.

FUTURE

BMSCs have been used for nearly last 3–4 decade to treat leukaemias, lymphomas, inborn errors of metabolism, autoimmune diseases such as Crohn disease, multiple sclerosis and rheumatoid arthritis. Intracoronary transplantation of stem cells following myocardial infarction has however shown only a modest benefit. Stem cell is not used in routine clinical practice in any part of the world. Most of the work in this area is done under research protocol that is scrutinised, monitored and then approved by the ethical committees of the respective research centres or hospitals or institutes. Currently, stem cell therapies are proceeding into placebo-controlled and double-blind clinical trials. These scrupulously designed trials shall lead us to exciting new therapeutic approach to human diseases.

Summary

Stem Cells are undifferentiated cells having an ability to divide and give rise to different types of cells when put under appropriate conditions.

Stem cells are of two types:

i) Embryonic stem cells (ESCs)
ii) Adult stem cells

They can also be classified as:

i) Totipotent
ii) Pluripotent

Embryonic Stem Cells can give rise to any tissue, e.g. cardiac muscle, liver, kidney, etc., and thus hold tremendous potential.

Adult Stem Cells are found among differentiated cells of a tissue or an organ. They help in repairs. They are found in bone marrow, peripheral blood, dental pulp, epithelia of gastrointestinal tract (GIT), etc.

Stem Cell Therapy: It is used to treat leukaemia and other blood disorders.

i) Bone marrow stem cell therapy (bone marrow transplant)
ii) Peripheral blood stem cells (PBSCs) therapy

Continued

Summary—cont'd

Umbilical Cord Blood Stem Cell Transplant: The potential that cord blood holds has led to "Cord Blood Banking". Cord blood banking has much greater prospects. However, the use of embryonic stem cells in research and therapeutics is under lens because of ethical, legal and social issues involved in it.

QUESTION YOURSELF*

1. What are stem cells?
2. What are the sources of stem cells?
3. What are applications of stem cells?
4. What are the ethical issues of stem cell therapy?

*See page 267 for Answers

Epigenetics 17

LEARNING OBJECTIVES

At the end of the chapter the student should be able to understand

- What is epigenetics?
- What is/are the mechanism/s involved in epigenetics?
- What is genomic imprinting?
- What are the applications of epigenetics?

KEY WORDS

Epigenetics, Genomic imprinting, Uniparental disomy, Histone modifications, Prader–Willi syndrome, Angelman syndrome

The word "epigenetics" is derived from the Greek word "Epi", which means over/around. So, epigenetics means something above or in addition to genetics, which normally follows the traditional mode of inheritance. Epigenetics describes phenotypic changes without altering the DNA sequence. It involves alterations in the "expression of gene" which eventually leads to phenotypic changes.

The epigenetic changes occur at cellular level and end up in physiological phenotypic traits. The extraneous factors, i.e. environmental factors seem to cause this epigenetic phenomenon. It also involves functional changes in the genome without marking structural alterations.

The best example of this is in the form of

1. DNA methylation
2. Histone modification

Both of these alter the gene expression. It is controlled by "repressor proteins" and the silencer regions of the DNA sequence. Epigenetic changes are observed in the stem cells (pluripotent cells) of the embryo during morphogenesis, the critical period of human development (2nd to 8th week of gestation). A unicellular stage of human development, i.e "zygote", by its repeated divisions (cleavage) forms highly differentiated cells of different types such as neurons, muscle cells, bone cells, various epithelial and connective tissue cells.

In order to understand epigenetics we will try to learn it under the following heads:

1. Definition
2. Its molecular basis and mechanisms
3. Its effects on development and transgenerational (heritable) effects
4. Applications

DEFINITION

A British embryologist coined the term "epigenetics" in 1942. However, it was done in the context of embryonic development, when differentiation of totipotent cells was being referred. In those days precise role of genes was not known. Subsequently, many researchers defined the term epigenetics differently. The widely accepted definition is as follows.

Epigenetics

This refers to both heritable gene activity and expression and also the long-term changes in the transcriptional potential of a cell that is not essentially heritable. In 2008, at Cold Spring Harbor meeting the researchers defined the term epigenetic trait.

Epigenetic Trait

This refers to heritable phenotype resulting from the changes in chromosome without alterations in the DNA sequence.

This has led to number of related terms like

- Genome—Epigenome
- Genomics—Epigenomics
- Genetic code—Epigenetic code

Molecular Basis of Mechanisms

Epigenetic alterations modify activation of certain genes without disturbing the DNA sequence. There occurs "activation" of some genes and "silencing" of some genes. Majority of these changes are preserved during cell division and restrict within the lifetime of that organism. However, some of these epigenetic alterations are transmitted to the next generation. They form what is called *transgenerational epigenetic inheritance.*

This occurs when gene silencing has involved either sperm or secondary oocyte. This leads us to a "*genomic imprinting*".

In Mendelian inheritance, maternal and paternal copies of gene (in allelic pair) are similar. However, some genes inherited from the father may have different behaviour from those inherited from the mother, though both have the same DNA sequence. This occurs due to heritable but reversible modification of the gene during gametogenesis. This phenomenon is called "imprinting". If the paternal allele is rendered inactive and silenced it is called "paternal imprinting". Alternatively, if the maternal allele is rendered inactive and suppressed it is called "maternal imprinting". This can be evidenced in pedigree.

Uniparental Disomy

In this, one of the parents contributes both the homologous chromosomes; chromosome from the other parent is absent. If a person has maternal isodisomy of a chromosome bearing maternally imprinted gene (silenced/inactive), then the person will not have functionally active gene, though he/she has both copies of gene having normal DNA sequence, e.g. *Prader–Willi syndrome* and *Angelman syndrome* (Fig. 17.1).

Genesis of uniparental disomy being nondisjunction in meiosis of one of the parent leading to either maternal or paternal isodisomy. This puts both

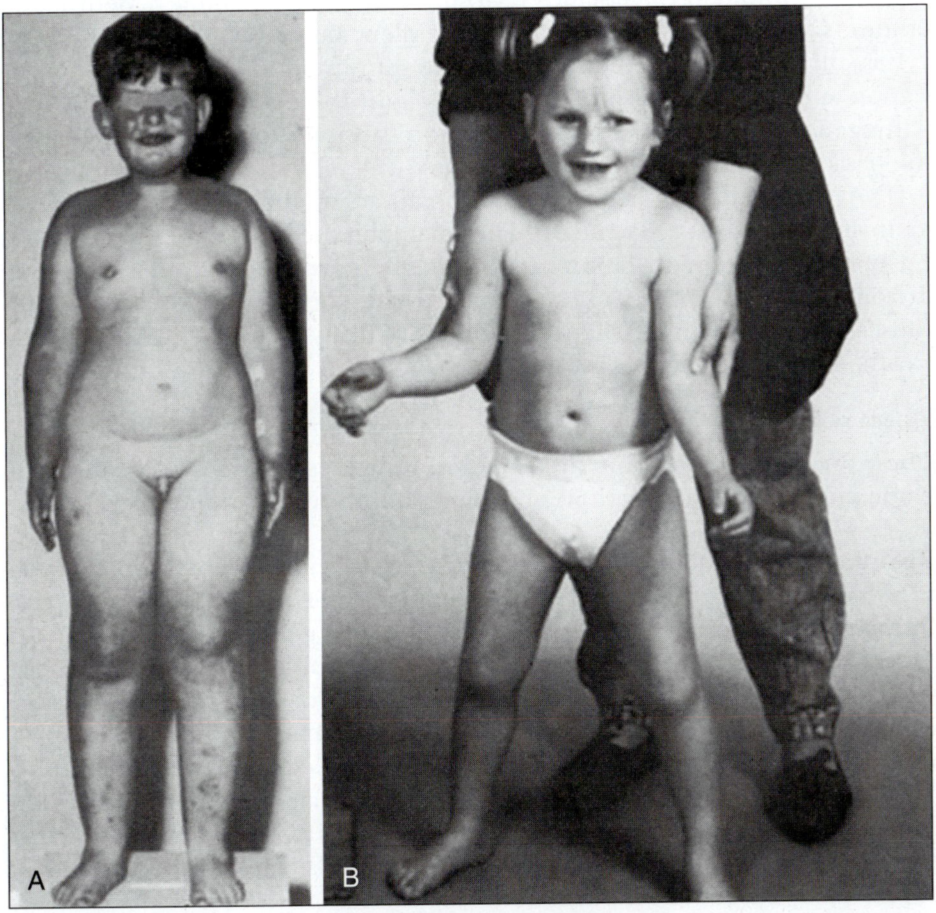

Figure 17.1 Photographs showing (A) Prader–Willi syndrome and (B) Angelman syndrome. (*Source*: From Nussbaum RL, McInnes RR, Willard HF. *Thompson and Thompson Genetics in Medicine*. 8th ed. Philadelphia: Elsevier, 2016.)

copies of chromosome pair in one gamete. When this fuses with normal gamete with one of copy of the chromosome, the zygote has three copies of this chromosome. Probably as a rescue mechanism one of the chromosomes is lost, and if the remaining two are from one parent (non-disjunctional gamete) then it leads to uniparental disomy.

Mechanism of Genomic Imprinting

It is one of epigenetic processes. It involves

1. DNA methylation
2. Histone modification

DNA methylation and CPG islands

In many genes at their 5′ end of the first exon, there is high concentration of CG sequences. About 3% of the cytosines in human DNA are methylated (addition of methyl group at 5′ carbon atom) forming 5-methylcytosine. Majority of them are found in CpG islands (cytosine–phosphate–guanine). Methylation of these CpG islands is associated with the silencing of the gene (imprinting).

It is still not clear how methylation stops transcription. An abnormal methylation and demethylation of genes is thought to be an important factor in pathogenesis of cancers. DNA methylation causing silencing of the genes is the basis of lyonisation of X chromosome.

DNA methylation is heritable and is caused by following enzymes DNA methyltransferases—2, 3a and 3b. They maintain the DNA methylation status.

Defects in DNA methylation can lead to genetic disorders and is even seen in some cancers. Drug 5-azacytidine prevents methylation and eventually causes expression of the silenced gene. It is likely to be accepted as a new treatment modality.

Histone modifications

These are unstructured and are called histone tails. These modifications include methylation, acetylation, phosphorylation, ribosylation.

RNA transcripts: At times, the gene after being "turned on" transcribes a product which maintains the activity of that gene, e.g. Hnf4 and MyoD. They increase transcription of number of liver and muscle-specific genes, respectively.

Micro RNA (miRNA): These are members of non-coding RNAs with size ranging from 17 to 25 nucleotides. They control and regulate number of biological functions in animal and plant kingdoms. In human, almost 200 miRNAs have been discovered. Each miRNA in a cell downregulates about 100–200 messenger RNAs. It is observed that about 60% of the human protein coding genes are regulated by miRNAs and nearly 50% of these miRNAs are associated with the CpG islands. These miRNAs may be repressed by epigenetic methylation.

Messenger RNA (mRNA): It is observed that methylation of mRNA plays a critical role in human energy homeostasis. An obesity-associated FTO gene is able to demethylate N6-methyladenosine in the RNA.

sRNAs: These are non-coding fragments found in bacteria. They regulate gene expression responsible for the virulence of bacteria. This is of prime importance in development of newer drug formulation against the drug-resistant bacteria. The sRNA–gene maps, based on their target in the microbial genomes, are being developed. This will certainly help the pharmaceutical industry to develop effective drug formulations.

Prions: These are infectious proteins which are capable of forming infectious conformational state. They also have the ability to catalytically convert native state version of the same protein. In this context, they are epigenetic agents which can bring about phenotypic change in the genome.

Commonly known are fungal prions, these lead to heritable infectious phenotype without modifications in genome. Couple of studies on yeast have led to the discovery of prions, PSI and URE3.

Nucleosome positioning: There are numerous nucleosomes in eukaryotic genome. Different tissues have different nucleosome positioning. This essentially determines differing gene expression as well as cell differentiation. The nucleosome positioning is heritable, and recently researchers have been able to correlate it to some epigenetic factors such as hydroxymethylation and methylation of DNA.

DNA damage: At times DNA damage can also cause epigenetic changes. It occurs frequently in natural course. Usually it is followed by DNA repair; however, the epigenetic change may remain as such. It is believed that normally DNA damage occurs with the frequency of 60,000 per cell in human body. It is to that tune and alarming proportion; however, it is duly compensated by DNA repair mechanism. An enzyme PARP-1 (Poly ADP-ribose-polymerase) and its product poly(ADP–ribose) (PAR) accumulate at the site of DNA damage. This directs recruitment and activation of chromatin remoulding protein ALC1. This helps in nucleosome remodelling. The nucleosome remodelling is believed to cause epigenetic silencing of DNA repair gene MLH1. Many chemicals have been identified to cause DNA damage, e.g. benzene, styrene, trichloromethylene and carbon tetrachloride.

Food: Among food ingredients, "isoflavones" can reduce DNA damage. Some food components increase the levels of DNA repair enzymes such as MGMT and MLH1.

Covalent modifications in DNA: They play a pivotal role in epigenetic inheritance. It may be through cytosine methylation involving DNA or the histone protein undergoing modification such as lysine acetylation; threonine and serine phosphorylation; and lysine and argentine methylation.

The epigenetic effects are seen due to heritable transcription states, which in turn cause change in phenotype due to alterations in the genes that are transcribed.

The gene expression is regulated by remodelling of chromatin. Chromatin comprises two things—DNA and histone protein. The remodelling of chromatin is accomplished through two mechanisms:

1. **Post-transitional modification:** This involves amino acids that form histone protein. This changes the shape of histone (bead). During

replication of DNA, it is possible that some modified histones may be carried into new copy of DNA. Then these histones may act as templates, initiating surrounding new histone to be shaped in a new manner. These modified histones continue lineage-specific transcription after cell division.

2. Another mechanism is **addition of methyl group to cytosine**, converting it into 5-methylcytosine. This, however, serves like regular cytosine and pairs with guanine. However, in some areas in genome which are highly methylated, they become transcriptionally less active. If this methylation involves germ cell line, it is passed on to the zygote. The heritability of methylation state depends on some enzymes, e.g. DNMT1, which have more affinity for 5-methylcytosine as compared to cytosine.

EFFECTS OF EPIGENETICS ON DEVELOPMENT

When we consider epigenesis and development together, it can be dealt by the following two ways:

1. **Predetermined epigenesis:** In this there is unidirectional effect in which the flow is structural development in DNA to functional maturation of the protein. It means the development is predetermined, i.e. already scripted and is thus predictable also.
2. **Probabilistic epigenesis:** It is a bilateral phenomenon between structure and function that is governed by the extraneous forces and experiences.

Transgenerational Epigenetic Inheritance

It has been widely studied in both plants and microbes. It is believed that the epigenetic mechanisms, e.g. DNA methylation and histone modifications, are heritable with the "principle of natural selection". So, basically these epigenetic mechanisms are a part of evolution relating to the cell differentiation.

Epigenesis in Bacteria

Bacteria make use of postreplicative DNA methylation to epigenetic control, DNA–protein interactions. In bacteria, particularly, DNA–adenine methylation is used as a signal and is important in bacterial virulence, e.g. *Escherichia coli, Salmonella, Brucella.* In these, adenine methylation regulates the cell cycle and provides signal for DNA replication.

Applications in Medicine

Epigenetics has a wide range of applications in medicine and health care system. NIH took up a 5-year programme to promote epigenetic research. Epigenetics can explain the underlying mechanism in human development, the process of aging, cardiac diseases, mental illness, cancer and numerous such conditions.

Twin Study

A comparison of monozygotic/identical twins provides one of the best models to observe the effects of environmental epigenetics. In the early part of life the effect of environment is minimal with very little exposure, but as these twins grow up, remarkable differences are seen in them. The genomic difference is seen in distribution of 5-methylcytosine and histone acetylation involving H3 and H4.

Dizygotic twins

They too show evidence of epigenetic influences.

In one of the study on twins (both mono- and dizygotic), analysis of DNA methylation status of about 6000 unique genomic regions revealed epigenetic similarity at blastocyst, which possibly contributes to phenotypic similarity in monozygotic twins. This also indicates that microenvironment at early embryonic stages has a significant role in establishing epigenetic markers.

Genetic Disorders

It is clear that epigenetics has a role in genesis of some of the genetic disorders, e.g. *Prader–Willi syndrome* and *Angelman syndrome.* Genes on chromosome *15q* are maternally imprinted.

Prader–Willi syndrome

Maternal disomy or deletion on paternal chromosome leads to *Prader–Willi syndrome.* Clinically, it presents with obesity, mental retardation, behavioural problems, hypogonadism. Face typically shows small forehead and almond-shaped eyes; hands and feet are small (Fig. 17.1a).

Angelman syndrome (Fig. 17.1b)

Gene associated with this syndrome is paternally imprinted. So deletion of gene on the maternal chromosome 15 or paternal disomy leads to Angelman syndrome. Clinically, it presents with characteristic face, mental retardation, puppet-like movements of hands, microcephaly. Epilepsy is also present in some cases.

Cytogenetic Findings

In both Prader–Willi syndrome and Angelman syndrome, cytogenetic findings show 15q deletion, i.e. deletion involving long arm of 15 chromosome and the bands missing between q11 and q13.

Cancer

Epigenetic mechanisms in different types of cancers are different. There may be epigenetic alterations involving DNA repair genes or the cell cycle controlling genes. The epigenetic alterations are important in cellular transformation into cancer. A further research on these lines can help in early detection, effective management and even prevention of cancer. This is duly addressed under what is called Cancer Epigenetics.

Diabetes

In diabetic patients, wound healing is a challenging task for the health care providers. The pathophysiology involving this reveals that hyperglycaemic environment induces/imprints such changes at genomic level which primes macrophages towards proinflammatory state and they do not subserve their *prohealing tendency*. Thus occurs *macrophage polarisation,* altering their normal role leading to problem in wound healing, which is one of the complications in the diabetic patients.

Psychology

Early-life stress (ELS): A follow-up study of 1000 subjects from preschool to adulthood revealed that those who carried one or two copies of short allele of serotonin transporter promoter polymorphism showed higher incidence of adult depression and suicidal tendency.

Addiction: It is observed that repeated exposure to the addictive stimulus, e.g. cocaine, morphine, gambling, through neuroepigenetic mechanism and transcriptional mechanism leads to addictive phenotype. In this, even transgenerational epigenetic inheritance has been observed.

Fear conditioning: In experimental animals (mice), the fear conditioning was also transmitted up to two generations. This was transmitted epigenetically through site-specific DNA methylation.

Anxiety: Anxiety-related phenotypes also show transgenerational epigenetic inheritance. Paternal stress-induced traits are transmitted through small non-coding RNA signals via sperm.

Depression epigenetic inheritance: It takes similar route as anxiety, i.e. through sperm via small non-coding RNA; it is transmitted from father to offspring.

Pharmacology

Some drugs or chemicals have epigenetic effect.

- **Opioid drugs:** They increase expression of genes associated with addictive phenotypes.
- **Beta-lactam antibiotics:** They can alter glutamate receptor activity and action of cyclosporine on multiple transcription factors.
- **Lithium:** It can cause autophagy of aberrant proteins.

Summary

- **Epigenetics:** It is something above/around or in addition to genetics which follows traditional modes of inheritance. It is defined as stably heritable phenotype resulting from changes in chromosome without alterations in the DNA sequence.
- **Molecular Basis:** Molecular basis of epigenetic changes involves activation of some genes and silencing of some genes without change in the DNA sequence.
- **Genomic Imprinting:** In Mendelian inheritance the maternal and the paternal copies of the gene are similar, but they may have different behaviour. It involves a heritable but reversible modification of the gene during gametogenesis. This is called "imprinting". If the paternal allele is rendered inactive/silenced, it is called **paternal imprinting**. Alternatively, if the maternal allele is silenced, it is called **maternal imprinting**.
- **Uniparental Disomy:** In this, one of the parents contributes both the homologous chromosomes; the chromosome from the other parent is absent. It is caused by non-disjunction during meiosis. If the person has maternal disomy of the chromosome bearing imprinted/silenced gene, then the person will not have functionally active gene, though DNA sequence in both the genes is normal, e.g. Prader–Willi syndrome.
- **Mechanism of Genomic Imprinting:** It involves DNA methylation and histone modifications. Association of CpG (cytosine–phosphate–guanine) islands at the 5′ end of the gene causes silencing of the gene, i.e. imprinting.
- **Histone Modifications:** They may occur through methylation, acetylation, ribosylation, etc.
- **Micro RNA (miRNA):** It is observed that about 60% of human protein coding genes are regulated by miRNA and nearly 50% of the miRNA are located in association with the CpG islands. These may be repressed by epigenetic methylation.
- **sRNA:** These are non-coding fragments found in bacteria. They regulate genes responsible for virulence of the bacteria. Obviously, these are important for new drug development. sRNA–gene maps are being developed in R&D wings of the pharmaceutical industry.
- **Prions:** Study on yeast has led to the discovery of prions PS1 and URE3. Prions are infectious proteins. They can cause heritable infectious phenotype without altering the genome.
- **Nucleosome Positioning:** They are different in different tissues and have an effect on differing gene expression and cellular differentiation. Nucleosome positioning is heritable and is related to epigenetic factor such as hydroxymethylation and DNA methylation.
- **DNA Damage:** It occurs with an alarming frequency of about 60,000 per cell, but is duly compensated by DNA repair mechanism. It eventually leads to nucleosome remoulding, which causes epigenetic silencing of DNA repair gene MLH1. DNA damage is caused by many chemicals, e.g. benzene, trichloromethylene, styrene, carbon tetrachloride.
- **DNA Modification:** Covalent DNA modifications have a pivotal role in the epigenetic inheritance. This may be through cytosine methylation or histone modifications. The epigenetic effect of it is seen due to heritable transcription, which eventually changes the phenotype.
- **Effect of Epigenetics on the Development:** It can be either of the following:
  1. **Predetermined epigenesis:** In this the flow is unidirectional, i.e. structural development of DNA to the functional maturation of protein.
  2. **Probabilistic epigenesis:** This has a bilateral flow between the structure and function, which is governed by extraneous forces and experiences.

Continued

Summary—cont'd

- **Transgenerational Epigenetic Inheritance:** It has been widely studied in plants and bacteria (microbes). Epigenetic mechanism remains the same, DNA methylation and histone modification. Basically, these epigenetic mechanisms are a part of evolution with the principle of "natural selection".
- **Epigenesis in Bacteria:** In bacteria, DNA adenine methylation is used as a signal and is of significance in bacterial virulence.
- **Applications in Medicine:** Epigenetics can explain mechanism of human development, aging process and pathogenesis of many diseases.
- **Twin Study:** In both monozygotic and dizygotic twins, effects of environmental epigenetics was studied. The study indicated that the microenvironment at the early embryonic stages has a significant role in establishing epigenetic markers.
- **Prader–Willi Syndrome:** The maternal disomy with the maternal imprinting involving chromosome 15q leads to Prader–Willy syndrome. It presents with obesity, mental retardation, behavioural problems. The face shows small forehead and almond-shaped eyes. Hands and feet are small.
- **Angelman Syndrome:** In this, there is paternal disomy of chromosome 15 with paternal imprinting. It is characterised by typical face, mental retardation, puppet-like movements of hands and microcephaly. Epilepsy may be present in some cases. Cytogenetic analysis in both the above syndromes reveals 15q deletion between q11 and q13 bands.
- **Cancer:** Epigenetic mechanisms are different in different types of cancers. They involve DNA repair genes and cell cycle controlling genes. Research in this field can help in early detection, effective management and even prevention of cancer. Ii is dealt with under Cancer Epigenetics.
- **Diabetes:** In the diabetic patients, non-healing ulcers pose a challenge. The hyperglycaemic environment imprints/induces changes at genomic level, thus priming macrophages to proinflammatory state leaving their natural tendency of "prohealing".
- **Psychology:** Early-life stress, addiction, fear conditioning and depression, all have epigenetic phenomenon involved in their genesis.
- **Pharmacology:** Some drugs and chemicals have epigenetic effect. These include opioid drugs, beta-lactam antibiotics, lithium.

QUESTION YOURSELF*

1. Define epigenetics.
2. Name two basic epigenetic mechanisms involving DNA.
3. What are CpG islands?
4. Explain genomic printing with examples.

*See page 268 for Answers.

Glossary

A

A An abbreviation for adenine.

Acentric A chromosome fragment without a centromere.

Acrocentric A chromosome having centromere at one end. Such chromosomes have satellited short arms carrying genes for rRNA.

Adenine It is a purine base in DNA and RNA.

Alleles They are alternative forms of gene at the same locus on homologous chromosomes. When there are more than two alleles at a given locus, they are called multiple alleles.

Allograft A graft where both donor and host belong to the same species but are not genetically identical.

Amber codon It is one of the three stop codons (UAG).

Amino acid An organic compound having both the carboxyl ($-COOH$) and amino ($-NH_2$) groups.

Amniocentesis A procedure by which amniotic fluid is obtained for prenatal diagnosis.

Anaphase The stage of cell division in which chromosomes migrate to opposite poles of the cell.

Aneuploid A chromosome number that is not an exact multiple of haploid number, i.e. $2n + 1$ or $2n - 1$, where n denotes haploid number of chromosomes.

Antibody It is an immunoglobulin produced in response to an antigenic stimulus and reacts specifically with the same antigen.

Anticipation This denotes an earlier onset and increased severity of some diseases in successive generations. It is believed to be an outcome of bias of ascertainment.

Anti-codon It is the complementary triplet of the tRNA that binds it with a particular amino acid.

Antigen A macromolecule that evokes antibody production by immunocompetent cells and specifically reacts with the same antibody.

Antigen-binding fragment (Fab) The part of antibody molecule that binds with the antigen.

Anti-parallel Refers to orientation of the two strands of DNA, one running in $5'$ to $3'$ direction and the other running in $3'$ to $5'$ direction.

Apoptosis Programmed cell death of a developing tissue or an organ of the body.

Ascertainment The method of selection of families for genetic study.

Association An occurrence of an allele in a group of patients more often than can be accounted for by chance.

Assortative mating The preferential selection of a mate with particular genotype.

Assortment It is random distribution of maternal and paternal chromosomes during gametogenesis. This also permits independent assortment of non-allelic genes to the gametes.
Autograft It is graft of the host's own tissue.
Autoradiography It is the procedure by which radioactively labelled molecule/s can be detected on an X-ray film.
Autosome Any chromosome other than sex chromosomes. There are 22 pairs of autosomes in humans.

B

B cells These are small lymphocytes, producing antibodies in response to antigenic stimulus.
Bacteriophage A virus that infects bacteria.
Balanced translocation A structural rearrangement of chromosomes in which genetic material is exchanged between two non-homologous chromosomes without loss or gain of the chromosome material.
Banding Procedure of staining chromosomes to visualise typical pattern of cross bands.
Barr body It is condensed inactive X-chromosome seen in a female somatic cell.
Base Refers to nitrogenous bases in nucleic acids, DNA and RNA (A—adenine, C—cytosine, G—guanine, T—thymine and U—uracil).
Base pair (bp) In DNA, complementary base A pairs with T and C pairs with G.
Bayes' theorem It states that combining the prior and conditional probabilities of certain events or the results of specific tests offers a joint probability in order to derive the posterior or relative probability.
Bias of ascertainment It is an artefact that must be taken into account during family studies while looking at segregation ratios, caused by families coming to attention because they have affected person/s.
Bivalent A pair of synapsed homologous chromosomes seen at metaphase of the first meiotic division.
Blood chimaera A mixture of the cells of different genetic origin present in twins.
Blood group Refers to system of red cell antigens.
Break point cluster (bcr) The region of chromosome 22 involved in Philadelphia translocation in chronic myeloid leukaemia.
Burden In genetics, it denotes the total impact of a genetic disorder in the patient, his family and the society.

C

C An abbreviation for cytosine.
Cancer family syndrome The term describes the clustering of particular types of cancers in certain families. It is thought that different types of cancer could be due to a single dominant gene, for example, Lynch type II.
Candidate gene A gene whose function suggests that it is probably the cause of a genetic disease.
Carrier A person who is heterozygous for a normal gene and an abnormal gene that does not express phenotypically but can be detected by specific tests.

CAT box It is non-coding, promoter sequence about 70–80 bp upstream from the site of initiation of transcription.
Complementary DNA (c-DNA) It is a single-stranded DNA and is transcribed from a specific RNA by an enzyme reverse transcriptase.
Cellular oncogene See protooncogene.
Central dogma The concept that the genetic information is transmitted from DNA to RNA to protein.
Centimorgan (cM) Also called map unit, it is used in linkage and is equivalent to 1% recombination.
Centric fusion The fusion of the centromeres of two acrocentric chromosomes giving rise to Robertsonian translocation.
Centriole A pair of cell organelles forming the points of focus of the spindle during cell division. They migrate to opposite poles of the cell during cell division.
Centromere It is also called kinetochore. It is the point at which the two chromatids of a chromosome are attached.
CFTR Stands for *cystic fibrosis transmembrane conductance regulator.* It is the gene product of cystic fibrosis gene, essential for chloride transport and mucus secretion.
Chiasma (Chiasma-cross) It denotes cross configuration of chromatids of the homologous chromosomes during the first meiotic division.
Chimaera An individual with two genetically different cell populations derived from different zygotes.
Chorion villous biopsy A procedure to obtain chorionic villous sample for prenatal diagnosis around 9–12 weeks under ultrasound control.
Chromatid During cell division each chromosome appears to be constituted by two parallel strands called chromatids held together by the centromere.
Chromatin The nucleoprotein fibres constituting the chromosomes.
Chromosome jumping or linking It is a technique of chromosome mapping. It involves circularisation of DNA fragments by restriction enzyme digestion in the presence of a plasmid sequence cut by the same restriction enzyme followed by digestion with second restriction enzyme that does not cleave within the plasmid sequence.
Chromosome mapping Assigning a gene to a specific chromosome or to particular region of a chromosome.
Chromosome painting The hybridisation in situ of fluorescent-labelled probes to chromosome preparation allowing identification of a particular chromosome.
Chromomeres They are densely coiled regions of chromatin on a chromosome giving the latter a beaded appearance, especially evident at meiotic prophase.
Chromosomal aberration A structural or numerical abnormality of chromosomes.
Class switching The term describes the normal change in antibody class from IgM to IgG in the immune response.
Cistron The smallest unit of genetic material that specifies synthesis of a particular polypeptide.
Clone A cell line derived from successive mitosis of a single diploid ancestral cell.
Codominant When both alleles of a pair are expressed in heterozygote state, the alleles are said to be codominant.

Codon A triplet of three nitrogenous bases that codes for one amino acid.

Coefficient of inbreeding (F) The probability that a person has received both alleles of a pair from an identical ancestral source.

Collinearity Term denotes the parallel relationship between base sequence of the DNA or mRNA and the amino acid sequence in corresponding polypeptide.

Complement There are about 10 serum proteins in humans that on activation interact in sequence to destroy cellular antigens.

Concordant When both members of a twin pair exhibit the same trait, they are called concordant.

Conditional probability Relates to tests or observations that can be used to modify prior probabilities in Bayesian calculations in risk estimation.

Congenital Refers to an abnormality present at birth; it may or may not be genetic in nature.

Consanguinity A relationship by descent through a common ancestor.

Cosmid A plasmid in which the maximum DNA has been removed to permit largest possible insert for cloning, but still has DNA essential for in vitro packaging into a phage particle.

Consultant An individual who gives genetic counselling.

Contigs Contiguous or overlapping DNA clones.

Control gene A gene that can turn other genes "on or off".

Cor pulmonale It is right-sided heart failure, as a sequel to a severe lung disease, e.g. pulmonary infection in cystic fibrosis.

CRISPR/Cas9 It is a recently discovered gene editing tool.

Crossing over Exchange of genetic material between two chromosomes of a pair such as in chiasmata formation in diplotene stage.

Cytosine A pyrimidine base in DNA and RNA.

Cytogenetics It is a branch of genetics that deals with the study of chromosomes.

Cytoplasmic inheritance Refers to transmission of a trait through the genes present in cytoplasmic organelles such as mitochondria.

D

Daltonism In the past, X-linked inheritance was called Daltonism, after John Dalton who noted this mode of inheritance in colour blindness.

Deletion A chromosomal aberration in which a part of chromosome is lost.

Denaturation Refers to conversion of double-stranded DNA into single-stranded DNA by destruction of bonds between base pairs.

Dermatoglyphics The study of patterns of skin ridges of fingers, palms and soles.

Developmental field It consists of apparently unrelated embryonic primordia that react together against a single genetic or environmental insult, thus producing a particular pattern of congenital malformations.

Dicentric An abnormal chromosome with two centromeres.

Dictyotene The stage of the first meiotic division. Human oocyte remains in this stage from prenatal life until ovulation.

Diploid The number of chromosomes in somatic cells of an individual. It is double the number found in gametes. In humans, diploid number is 46 ($2n$).

Discordant When only one member of a twin pair shows a particular trait and other does not, they are said to be discordant.

Dispermy Fertilisation by two sperms, of a duplicated egg nucleus or of egg and polar body.
Diversity region They are DNA sequences coding for segments of the hypervariable regions of antibody molecule.
Dizygotic twins, fraternal twins Twins produced by fertilisation of two separate ova by two different sperms.
DNA (deoxyribonucleic acid) Nucleic acid in chromosome that stores and transmits genetic information.
DNA cloning Production of many identical copies of a DNA fragment.
DNA fingerprinting The pattern of hypervariable tandem DNA repeats of a core sequence that is unique to a person.
DNA library It is the collection of recombinant DNA from a particular source, e.g. genomic or cDNA.
DNA ligase It is an enzyme that catalyses formation of a phosphodiester bond between a 3′-hydroxyl and a 5′-phosphate group in DNA, thus joining two DNA fragments.
DNA mapping It denotes the physical relationship of flanking DNA sequence polymorphisms and the detailed structure of a gene.
DNA probes These are DNA sequences that are labelled radioactively and are used to identify a gene, e.g. genomic probe.
Dominant A trait that expresses even in heterozygote state for a particular gene.
Dosage compensation It refers to the fact that the amount of product of an X-linked gene is equal irrespective of the number of X chromosomes.
Double-minute chromosomes They are amplified DNA sequences in tumour cells that occur as small extra chromosomes as found in neuroblastoma.
Drift The fluctuations in gene frequencies that tend to occur in small isolated populations.
Duplication A chromosomal aberration in which a part of chromosome is duplicated.
Dystrophin It is a protein, the gene product of Duchenne muscular dystrophy gene (DMD gene).

E

Ecogenetics It is the study of genetically determined differences in susceptibility to the action of physical, chemical or an infectious agent in the environment.
Endoreduplication It refers to duplication of a haploid sperm chromosome set.
Empiric risk Probability of recurrence of a trait in a family based on previous experience and observation.
Endonucleases Enzymes that can cleave bonds in DNA or RNA strand.
Enhancer A DNA sequence that promotes transcription of the related gene.
Enzyme A protein that serves as a catalyst in biochemical reactions.
Epidermal growth factor (EGF) It is a growth factor that stimulates a variety of cell types inclusive of epidermal cells.
Epigenetics It refers to both heritable gene activity and expression and also the long term changes in the transcriptional potential of a cell that is not essentially heritable
Euchromatin Represents genetically active regions of the chromosomes.

Eugenics A branch of genetics that promotes the improvement of hereditary qualities of a race/species.
Eukaryote An organism in which cells have a nucleus and a nuclear membrane.
Exon A segment of gene that is represented in mRNA product and codes for protein.
Expressivity Refers to severity of the expression of a particular gene.

F

Fab An antigen-binding fragment of an antibody molecule produced by digestion with papain.
FACS Stands for *fluorescent activated cell sorting*. It is a technique in which chromosomes are stained with fluorescent stain that selectively binds to DNA. The difference in fluorescence of various chromosomes allows their separation using special laser.
First filial generation (F_1) The first generation progeny of a mating.
Familial Any trait that is more frequent in relatives of an affected person than in the general population.
Filial Offspring.
First-degree relatives Close relatives, i.e. parents, offsprings, sibs. They share an average 50% of their genes.
FISH Stands for *fluorescent in situ hybridisation*. The technique in which single stranded DNA with a fluorescent label is hybridised with its complementary target sequence in the chromosome, permitting its visualisation under UV light.
Fitness Refers to biological fitness in terms of number of offsprings who reach reproductive age. Fitness is taken as unity (100%) if the person and his/her spouse have two such offsprings.
Five-prime (5′) end The end of a DNA or RNA strand having a free 5′ phosphate group.
Flanking sequences Nucleotide sequences preceding or following the transcribed region of gene.
Flow cytometry See fluorescent-activated cell sorting.
Foetoscopy A procedure of direct visualisation of foetus for prenatal diagnosis.
Forme fruste A mild expression of a trait, bears no clinical significance.
Founder effect Refers to the high frequency of a mutant gene in the population founded by a small ancestral group when one of the founders was a carrier of the mutant gene.
Frameshift mutations The mutations such as deletions or insertions, which change the reading frame of the codon triplets.

G

G An abbreviation for guanine.
G bands They are dark and light bands seen in chromosomes after treatment with trypsin followed by Giemsa stain.
Gamete A germ cell (ovum or sperm) having haploid number of chromosomes.
Gene A part of DNA molecule that directs synthesis of a polypeptide chain or RNA molecule. It consists of many codons.
Gene amplification Production of multiple copies of certain genes in tumour cells; evidenced by homogeneously staining regions (HSRs) and double minute chromosomes.

Gene flow Diffusion of genes from one population to another, through migration and mating.
Gene map Represents human karyotype showing chromosomal localisation of the genes.
Gene pool Total genes present at a given locus in the population.
Gene therapy Management of an inherited disease by addition or insertion of a normal gene/s.
Genetic code Triplet of bases that specifies amino acids.
Genetic counselling Deals with providing information to patients and the relatives at the risk of a genetic disorder, the consequences of the disorder, the probability of recurrence and the ways by which it may be prevented or mitigated.
Genetic death Failure of a mutant gene to be passed on to the next generation because of its seriously damaging phenotypic effects.
Genetic lethal Refers to the gene or genetically determined trait that leads to failure of reproduction in an affected individual.
Genetic load The sum total of all kinds of harmful alleles in a population.
Genetic marker A trait can be used as a genetic marker in study of individuals, families or populations provided that the trait is genetically determined, has a simple pattern of inheritance, can be classified accurately and has variations common enough to permit it to be labelled as genetic polymorphism.
Genetic screening The screening tests in population designed to identify individuals at risk of having a specific disorder or are likely to produce an offspring with such a disorder.
Genetic susceptibility An inherent predisposition to a disease that is due to complex interaction of effects of multiple genes, i.e. polygenic inheritance.
Genome The entire DNA complement in the cell.
Genotype The genetic constitution of an individual (genome).
Goldberg–Hogness box See CAT box.
Gray (Gy) 1 Gy is equivalent to 100 rad.
Growth factor A substance that is essential in a culture medium to allow multiplication of the cells or the substances involved in growth of certain cell types, tissue or body part, e.g. fibroblast growth factor.

H

Haploid The number of chromosomes in a normal gamete. In humans, it is 23 (n).
Haplotype Refers to a group of alleles from closely linked loci. They are inherited as one unit. For example, HLA complex of four genes on each chromosome 6.
Hardy–Weinberg's law The law states that in large randomly mating population relative proportions of the different genotypes remain constant from one generation to another provided no evolutionary processes like migration, selection and drift are operating.
Hemizygous A term used to denote genes on X chromosome in males.
Heritability The proportion of the total variation of a character attributable to genetic as against environmental factors.
Hermaphrodite, intersex An individual with gonads of both male and female.
Heterochromatin Genetically inactive regions of the chromosomes.

Heterogametic sex The sex that produces gametes of two different types. In humans, male is heterogametic, as he produces X and Y bearing sperms.
Heteromorphism The heritable structural polymorphism in chromosomes.
Heterozygote Refers to an individual possessing two different alleles at a given locus on a pair of homologous chromosomes.
Histocompatibility A graft is accepted by a host if it is histocompatible; it means there exists an antigenic similarity between them.
Histone Type of protein associated with DNA in chromosomes, rich in lysine and arginine.
HIV Human immunodeficiency virus.
HLA Human leucocyte antigen. It is present on the surface of cells including lymphocytes.
HLA complex Refers to genes on chromosome 6, responsible for determining the cell-surface antigens. They are important in organ transplantation.
Holandric inheritance The pattern of inheritance of genes on Y chromosome. They pass from father to all his sons but to none of his daughters.
Homeobox A sequence of about 180 bp found to be conserved in different homeotic genes.
Homeotic gene Gene that controls development of a region of organism producing proteins that regulate gene expression by binding particular DNA sequences.
Homologous chromosomes A pair of chromosomes, one derived from each parent, having identical loci.
Homozygote An individual who has two identical alleles at a given locus on a pair of homologous chromosomes.
Human genome project A collaborative international effort to map and sequence the entire human genome.
H-Y antigen Histocompatibility antigen originally detected in mouse and earlier thought to be located on Y chromosome, now believed to be coded by a gene on chromosome 6 and controlled by a regulatory gene on Y chromosome.
Hybrid Refers to progeny of cross between two genetically different organisms.
Hybridoma The term denotes cells formed by fusion of mutant myeloma cells and other cells like spleen cells from an immunised mouse. The technique is applied to produce monoclonal antibodies.
Hypervariable DNA length polymorphism Refers to a number of different types of variation in DNA sequence that are highly polymorphic, e.g. variable number tandem repeats (VNTRs), microsatellites.

I

Idiogram An ideal representation of an object, e.g. an ideogram of a karyotype.
Immune reaction The reaction that occurs between an antigen and antibody.
Immunological tolerance Failure to react to antigen because of previous exposure to that antigen.
Inborn error A specific enzyme defect leading to a metabolic block and resulting in a genetically determined biochemical disorder.
Inbreeding The mating between closely related individuals.
Incidence The rate of occurrence of new cases, e.g. 1 in 3000 male births is affected by DMD.

Index case, proband The affected family member through whom the family is ascertained.
Inducer The molecule that interacts with a regulator protein and triggers transcription of gene.
Insertion Term denotes a structural chromosomal aberration involving addition of DNA sequence from non-homologous chromosomes.
Insertional mutagenesis The insertion of mutations at specific sites to ascertain the effect of these changes.
In situ hybridisation Hybridisation with a DNA probe, directly on a chromosome preparation.
Interphase Part of the cell cycle between two successive cell divisions
Intron The part of a gene that is initially transcribed into the primary transcript (hnRNA) but is then removed and is not present in mRNA.
Inversion A structural chromosomal abnormality in which a part of chromosome is inverted. It may be pericentric or paracentric.
In vitro In the laboratory, literally means "in glass".
In vivo In the cell, actually means "in the living organism".
Isochromosome An abnormal chromosome resulting from transverse division of centromere in which one arm is deleted. The isochromosome therefore has two arms of equal length bearing same genes.
Isograft A graft between two genetically identical individuals.
Isolate A small population group in which matings occur exclusively between members of the same population group.

J

Joint probability Denotes the product of the prior and the conditional probability for two events.

K

Karyotype The term denotes chromosome set. It is also used for photomicrograph of an individual's chromosomes arranged according to standard classification.
Kilobase (kb) One thousand bases in DNA or RNA.

L

Ligase An enzyme used to join DNA molecules.
Linkage The genes located close together on the same chromosome are said to be linked.
Linkage disequilibrium The tendency of two linked alleles to occur more frequently on the same chromosome than would be expected by chance.
Locus The site of a gene on a chromosome is called locus. Alternative forms of genes (alleles) may occupy the locus.
Locus control region (LCR) Located close to beta-like globin genes involved in tissue specificity and the timing of their expression in development.
Lod score A score of the relative likelihood of two loci being linked.
Long terminal repeat (LTR) Located at the ends of DNA synthesised by reverse transcriptase from the retroviral RNA and is involved in regulating viral expression.
Lymphokines Refers to a group of glycoproteins released from T lymphocytes.
Lyonisation (Lyon's hypothesis) Inactivation of genes on one of the X-chromosomes during the embryonic period in female mammalian somatic cells.

M

Manifesting heterozygote Denotes a female heterozygote for an X-linked disorder in whom the trait is expressed clinically with almost same severity as in hemizygous males.

Marker chromosome A small and structurally abnormal, extra chromosome.

Meiosis It is a special type of cell division occurring in germ cells and resulting in the formation of gametes with haploid number of chromosomes. There are two meiotic divisions, meiosis I and II. Chromosome number is reduced in meiosis I.

Messenger RNA (mRNA) It is transcribed from DNA and forms template for translation of protein.

Metaphase The stage of mitosis or meiosis in which chromosomes are condensed to their maximum capacity and are lined up at the equatorial plane of the cell.

Methaemoglobin Refers to haemoglobin molecule in which iron is oxidised.

MHC Stands for *major histocompatibility complex.* It is a locus with multiple genes, coding for histocompatibility antigens involved in organ transplantation and located on 6 P.

Minichromosomes Artificially constructed chromosomes possessing centromeric and telomeric elements that allow replication of foreign DNA as a separate entity.

Minigene A gene with majority of sequence removed but still remaining functional, e.g. dystrophin minigene.

Missense mutation A point mutation that results in an alteration in an amino acid specifying codon.

Mitochondrial DNA It is the circular DNA of mitochondria, a cytoplasmic structure. It is maternally inherited.

Mitosis The type of cell division that occurs in somatic cells. The daughter cells have the same chromosome complement as that of the parent cell.

Monoclonal antibodies See hybridoma.

Monosomy Refers to absence of one chromosome from a pair. For example, 45, X (Turner syndrome). Partial monosomy may also occur.

Monozygotic twins, identical twins The type of twin derived from a single fertilised ovum.

Mosaic An individual with at least two cell lines with different genotypes but derived from a single zygote.

Multifactorial Refers to the combination of multiple factors controlling inheritance, such as genetic factors and also the non-genetic (environmental) factors. It should be distinguished from polygenic.

Mutagen An agent that increases the mutation rate by changing DNA structure.

Mutation A permanent heritable alteration in genomic DNA sequence. When it involves a single gene it is called point mutation.

Mutation rate Denotes number of mutations at a given locus. It is expressed as mutations per gamete per locus per generation.

N

Neutral gene Refers to a gene that has no obvious effect on survival.

Nick A break in the sugar–phosphate backbone of RNA or DNA strand.

Non-disjunction Two members of a chromosome pair fail to separate (disjoin) during cell division. As a result both pass to the same daughter cell.

Nonsense mutation Mutation involving a chain termination codon.
Northern blotting The mRNA electrophoretic separation with subsequent transfer to nitrocellulose filter and radiolabelling.
Nucleosome The primary repeating unit of DNA structure in chromatin fibre.
Nucleotide Many nucleotides constitute nucleic acid. Each nucleotide comprises a nitrogenous base, a pentose sugar and a phosphate group.
Nucleus A structure within the cell that contains nucleolus and the chromosomes.

O

Obligate carrier Refers to a person who by pedigree analysis must carry a particular gene, e.g. the parents of a child with an autosomal recessive trait, the daughter of a man having an X-linked recessive disorder.
Ochre codon One of the three stop codons (UAA).
Oncogene The gene responsible for cancer.
Operator gene A gene that switches on an adjacent structural gene.
Operon It consists of an operator gene and the structural gene that it controls.

P

p Denotes (*i*) the short arm of a chromosome, (*ii*) frequency of more common allele of a pair in population genetics.
Packaging cell line Denotes cell line that has been infected by a retrovirus in which the provirus is genetically modified to lack the packaging sequence essential to produce infectious viruses.
Parthenogenesis The process of development of an organism from an unfertilised oocyte.
PCR Stand for *polymerase chain reaction.* It refers to a serial successive reaction using oligonucleotide primers and DNA polymerase to amplify desired DNA sequence.
PDGF Stands for *platelet derived growth factor.* It is derived from the platelets. It stimulates the growth of certain cells.
Pedigree A diagram of family history indicating normal and affected individuals, their relationship to the proband and their status with respect to a particular genetic disorder.
Penetrance The proportion of heterozygotes that express a trait, even though mildly.
PFGE Stands for *pulsed field gel electrophoresis.* It is a DNA analysis with electrophoresis to separate large DNA fragments (2 million bp size) obtained by DNA digestion using restriction enzymes.
Pharmacogenetics Deals with drug responses and their genetically controlled variations.
Phenocopy It is a copy of a phenotype. A condition that is due to environmental factors but mimics one that is genetic.
Philadelphia chromosome It is a structurally abnormal chromosome 22, found in bone marrow cells of many patients of chronic myelogenous leukaemia. It is formed by reciprocal translocation between the long arms of chromosomes 22 and 9.
Pleiotropy The phenomenon of a single gene presenting multiple effects.
Poly (A) tail A sequence of about 20–200 adenylic acid residues at 3′ end of mRNA, stabilising it and rendering it resistant to nuclease digestion.

Polygenic A trait determined by many genes at different loci, should be distinguished from multifactorial trait in which the environmental factors operate.
Polymorphism The occurrence in a population of two or more genetically determined forms, each with such frequencies that the rarest of them cannot be maintained by mutation alone.
Polypeptide An organic compound comprising a chain of amino acids held together by peptide bonds between amino group of one and the carboxyl group of the other.
Polyploid Any multiple of haploid number, other than diploid, such as $3n$, $4n$, etc.
Polysome A group of ribosomes (polyribosome) associated with a mRNA molecule.
Posterior probability The joint probability for a particular event divided by the sum of all possible joint probabilities.
Predictive testing Refers to presymptomatic testing of person/s at the risk of a disorder e.g. Huntington disease.
Preimplantation diagnosis Performed to detect presence of inherited disease in an in vitro fertilised conceptus before implantation.
Prior probability The initial probability of an event.
Proband See index case.
Probability The proportion of times an outcome occurs in a large series of events.
Probe A radioactive DNA or RNA sequence used to detect the complementary sequence by molecular hybridisation.
Processing Includes alterations in RNA that occur during transcription; these are splicing, capping and polyadenylation.
Prokaryotes These are lower organisms without a well-defined nucleus, e.g. bacterium.
Prophase The first visible stage of cell division in which chromosomes are seen as discrete structures. Subsequently they thicken and shorten.
Pseudogene A functionally inactive DNA sequence homologous with a known gene.
Pulsed field gel electrophoresis See PFGE.

Q

q Denotes (*i*) the long arm of a chromosome; (*ii*) frequency of rarer allele of a pair in population genetics.
Q bands The pattern of bright and dim cross-bands seen on chromosomes when observed under fluorescent light after quinacrine staining.

R

Rad The unit of measurement of any ionising radiation that is absorbed by the tissues. One rad is equal to 100 ergs of energy absorbed per gram of tissue.
Random mating, panmixis Selection of a mate without considering the genotype.
Recessive A trait that expresses only in homozygotes.
Recombinant DNA An artificially synthesised DNA in which a part of DNA sequence from one organism is inserted into the genome of another.
Recombination A crossing over between two linked loci.

Recombination fraction (ϑ) It is a measure of the distance, separating two loci and is determined by the likelihood/chance that a cross over will occur between them.
Recurrence risk It is the probability that a genetic disorder present in one or more members of a family shall recur in another member of the same or the following generation.
Reduction division The first meiotic division in which the chromosome number is reduced from diploid to haploid.
Regulator gene In accordance with the operon concept, a regulator gene synthesises a repressor substance that inhibits the action of operator gene.
rem Roentgen equivalent for man. It has the same biological effect as one rad of X-rays.
Restriction endonuclease An enzyme that cleaves DNA at a specific base sequence producing fragments of DNA, used in recombinant DNA technology.
Restriction fragment length polymorphism (RFLP) Polymorphism owing to the presence/absence of a specific restriction site.
Restriction map A linear arrangement of sites on DNA cleaved by various restriction enzymes.
Retrovirus RNA virus that replicates via conversion into DNA.
Reverse transcriptase An enzyme that catalyses the synthesis of DNA from RNA.
Ring chromosome A structural chromosomal aberration in which the terminal portion of both arms of a chromosome break off and the remaining chromosome forms a ring.
RNA Ribonucleic acid is found chiefly in nucleolus and ribosomes. It has pentose sugar ribose. RNAs are of the following classes—messenger RNA (mRNA), transfer RNA (tRNA), ribosomal RNA (rRNA) and viral RNA.
Robertsonian translocation A translocation involving two acrocentric chromosomes by fusion at the centromere and loss of their short arms.

S

Satellite A distal part of chromosome separate from the rest of the chromosome by a narrow stalk.
Secretor gene In humans, secretor gene is responsible for the secretion of ABO blood group antigens in saliva and other body fluids.
Segregation Refers to separation of alleles at meiosis, as a result two members of allelic pair pass to two different gametes.
Selection It refers to the operation of forces that determine the relative fitness of a genotype in population.
Sex chromatin, Barr body A darkly stained mass located at the periphery of the nucleus of a female mammalian cell during interphase. It represents an inactive X chromosome.
Sex chromosomes The chromosomes responsible for determination of sex, XX in females and XY in males.
Sex-determining region of Y See SRY.
Sex influenced A trait that is not X-linked but still expresses differently either in degree or in frequency, in males and females, e.g. congenital adrenal hyperplasia.
Sex limited A trait that is expressed in only one sex though it is not determined by an X-linked gene, e.g. precocious puberty in males.

Sex linkage Denotes genes carried on sex chromosomes. Since there are very few genes on Y chromosome, the term is often used synonymously for X-linkage.
sib Brother or sister.
Sievert (SV) It is equivalent to 100 rem.
Silencer It represses the gene expression.
Silent gene A gene that has no visible phenotypic effect.
Sister chromatid exchange An exchange of genetic material between the two chromatids of any particular chromosome, e.g. Bloom syndrome.
Solenoid Refers to a coil of wire wound around a hollow core. In cytogenetics, the term is used to describe the coiled structure into which nucleosomes are wound during chromatin condensation.
Somatic mutation A mutation that occurs in somatic cell rather than in the germ cell line.
Southern blot The technique for transferring DNA fragments from agarose gel to a nitrocellulose filter on which specific DNA fragment can be identified by hybridisation to radioactive probe. It is named after Edwin Southern, who devised the technique.
Spindle A structure that is responsible for the movement of the chromosomes during cell division. It consists of intracellular microtubules.
Splicing Involves joining of exons after the removal of introns in RNA during transcription.
Split gene A gene having one or more introns.
SRY Stands for *sex-determining region of Y*. It refers to the part of Y chromosome containing testis determining gene.
Stop codon One of the three triplets UAA, UAG and UGA, causes termination of protein synthesis.
Structural gene A gene coding for RNA or protein product other than regulator gene.
Switching An alteration in the type of α or β globin chains in prenatal period.
Syndrome The complex of symptoms and signs that are found together in any particular disorder.
Synthetic genes Two genes that occur at different loci on the same chromosome.

T

T An abbreviation for thymine.
T cells The small lymphocytes committed by the influence of the thymus and are responsible for cellular immunity.
TATA box (Hogness box) A conserved, non-coding DNA sequence about 30 bp upstream from the site of initiation of transcription.
Telophase The stage of cell division that commences when the daughter chromosomes reach the poles of the dividing cell and completes when the two daughter cells take an appearance of interphase cells.
Teratogen An agent that induces or increases the incidence of congenital malformations.
Termination/stop codon There are three termination/stop codons—UAG, UAA and UGA. Any one of them can terminate protein synthesis.
Three-prime (3′) end The end of the RNA or DNA strand having 3′ hydroxyl group.

Tissue typing Serological and cellular testing to ascertain histocompatibility before organ transplantation.
Trait A detectable phenotypic character.
Transcription The synthesis of mRNA or DNA template.
Transcription factors These include genes such as *Hox, Pax* and zinc finger genes that control RNA transcription by binding to specific DNA regulatory sequences. They form complexes that initiate transcription by RNA polymerase.
Transduction Involves transfer of DNA from the donor cell to the recipient cell/organism by recombination of the genetic material through a phage.
Transfection Refers to the transformation of a bacterial cell by infection with phage to produce infectious phage particles. Also insertion of foreign DNA into eukaryotic cells in the culture.
Transformation Denotes genetic recombination in bacteria, involving introduction of foreign DNA into the bacterium, its subsequent incorporation into the bacterial chromosome. Also the change of a normal cell into a malignant phenotype, e.g. normal cells infected by oncogenic viruses.
Translation Refers to the process by which genetic information along mRNA is translated into protein synthesis.
Translocation The transfer of genetic material from one chromosome to another non-homologous chromosome is translocation. If the two non-homologous chromosomes exchange genetic material, it is called reciprocal translocation. See also Robertsonian translocation.
Transposon Refers to the mobile genetic element and inserting a copy of itself at a new location in the genome.
Triplet codon In genetics, a unit of three bases in DNA or RNA that codes for an amino acid.
Triploid A cell having three haploid sets of chromosomes ($23 \times 3 = 69$ chromosomes).
Triradius In dermatoglyphics, the term denotes a point from which the dermal ridges course in three directions at angles of approximately 120°.
Trisomy Refers to a state of having three representatives of a given chromosome instead of normal two, e.g. Down syndrome or trisomy 21.

U

U An abbreviation for uracil.
Ultrasonography A procedure in which high frequency sound waves are used to delineate the outline of various internal structures.
Unifactorial Inheritance controlled by a single gene pair.
Utrophin Refers to a gene on chromosome 6 with homology to dystrophin gene.

V

Vector A plasmid or phage in which foreign DNA may be inserted for cloning.
Virions Infectious viral particles.

X

X-chromatin See Barr body.
Xenograft A graft from donor of one species to the host of different species.

Y

Yeast artificial chromosome (YAC) It is a plasmid-cloning vector that contains DNA sequences for centromere, telomere and autonomous chromosome replication sites, which enable cloning of large DNA fragments up to 2–3 million bp length.

Z

Zinc finger Finger-like projection formed by amino acids, positioned between two cystine residues that form a complex with a zinc ion. It is found in genes having important regulatory role in development.
Zoo blot Refers to Southern blot of DNA from different species used to look for evidence of conservation of DNA sequences during evolution.
Zygote A diploid cell resulting from union of male and female gamete (fertilisation).

Answers

Chapter 1: Historical Gleanings

1. c
2. b
3. d
4. d

Chapter 2: Cytogenetics

1. a
2. It is derived from "chroma" meaning colour and "soma" meaning body. The chromosomes appear like coloured (stained) rod shaped structures.
3. A cell that is capable of division undergoes a cyclical change throughout its life. This is called cell cycle.
4. Most of the cells in the body exist and function in an interphase.
5. The interphase consists of the following:
   a. G1 (Gap 1 phase).
   b. S (Synthesis phase): DNA is synthesised in this period.
   c. G2 (Gap 2 phase): a brief interval before commencement of mitosis.
6. These are:
   a. Formation of metaphase plate with chromosomes arranged at equatorial plane.
   b. Formation of spindle.
7. The centromere divides vertically and the paired chromatids disjoin to form daughter chromosomes.
8. Addition of colchicine (or its derivatives) inhibits spindle microtubule formation. This permits us to study metaphase chromosomes.
9. Somatic recombination is crossing over between homologous chromosomes in mitosis.
10. It involves crossing over between the sister chromatids of a single chromosome in mitosis.
11. In anaphase of meiosis I, without split of the centromere, two members of bivalent (homologous pair) disjoin.
12. Meiosis II differs from mitosis in two respects—(*i*) there is no DNA replication prior to this and (*ii*) the second meiotic division follows meiosis I without an interphase.
13. a
14. a
15. d

16. c
17. It is the process of union of mature male and female germ cells, resulting in formation of zygote.
18. c
19. (*i*) Metacentric, (*ii*) submetacentric, (*iii*) acrocentric, (*iv*) telocentric.
20. b
21. c
22. Secondary constrictions are evident on C banding in chromosomes 1, 9, 16 and long arm of Y chromosome.
23. Routine metaphase preparation shows total of about 200 bands; in high resolution banding, prometaphase chromosomes are studied and they exhibit about 800–1400 bands, making minor structural alterations evident.
24. In this, the somatic cells from two different species are fused together under favourable conditions and cultured.
25. A recent technique for chromosome analysis involving fluorescent activated cell sorting.
26. It is fluorescent in situ hybridisation.
27. Karyotyping is useful in (*i*) confirming clinical diagnosis, (*ii*) gene mapping, (*iii*) prognosis of disease, e.g. chronic myelogenous leukaemia, and (*iv*) prenatal diagnosis.
28. It is a small chromatin mass in nucleus of female somatic cell. It represents an inactive X chromosome.
29. The number of Barr bodies in a cell is $n - 1$ where n stands for number of X chromosomes, i.e. if the chromosome complement of a cell is 46, XX—number of Barr bodies is $2 - 1 = 1$; 47, XXX, $3 - 1 = 2$; 48, XXXX, $4 - 1 = 3$ Barr bodies.
30. Lyon's hypothesis states the following:
    a. In female somatic cell, only one X chromosome is active. The second is inactive and condensed to form sex chromatin.
    b. The inactivation occurs early in embryonic life.
    c. The inactivation is random but fixed.
31. It is achieved through DNA methylation.
32. The inactivation centre is believed to be in proximal part of long arm of X chromosome.

Chapter 3: Molecular Genetics

1. Operon is a unit that comprises an "operator gene" and "structural gene". The operator gene controls the action of structural gene and thus governs the amount of protein product produced.
2. It is defined as a change in sequence of genomic DNA.
3. It is also called as single-base substitution. In this, there is a single base change in DNA sequence. This alters the triplet code and causes replacement of amino acid.
4. It is true. Females are mosaic in respect to X chromosome. They possess two cell populations, one cell line with one X chromosome (X^m or X^p) active and the other with an alternative X active.

5. The nucleic acid consists of long chain of molecules called nucleotide. Each nucleotide molecule in turn consists of a nitrogenous base, a sugar moiety and phosphorous molecule.
6. b
7. a
8. It is a sequence of three bases that codes for one amino acid. The genetic information regarding protein synthesis is stored in DNA molecule in the form of triplet code.
9. They are as follows:
   a. hnRNA—heterogeneous RNA
   b. mRNA—messenger RNA
   c. tRNA—transfer RNA
   d. rRNA—ribosomal RNA
10. The primary product of transcription is a heterogeneous nuclear RNA molecule.
11. It is a process whereby information is transmitted from DNA to the messenger RNA.
12. It is a process of translating information from mRNA into protein synthesis.
13. It is a segment of DNA molecule possessing code for amino acid sequence of a polypeptide chain. It has exons (coding sequence), introns (non-coding sequence) and flanking regions serving as "start" and "stop" signal.
14. They are also called "transposons"; these are regions of DNA, which can jump to and fro within single chromosome or an adjacent one. The credit of their discovery goes to Barbara McClintock, an American geneticist.
15. These are DNA sequences that show a striking similarity with functional genes but are not transcribed. Probable reason being that their regulatory regions are altered by mutation.
16. The deletion or insertion of a base leads to an alteration in the reading frame of DNA strand and thus the amino acid sequence. This is called as frame shift mutation.
17. It is "pulsed field gel electrophoresis", which is used for DNA/gene mapping.
18. It is a DNA molecule obtained by combining DNA sequence from different organisms.
19. Plasmid is a circular chromosomal element in bacteria.
20. d
21. PCR means polymerase chain reaction. It is an in vitro method of synthesis of nucleic acid, wherein, a specific DNA segment is amplified rapidly without concomitant replication of the rest of the DNA.

Chapter 4: Chromosomal Aberrations

1. b
2. a

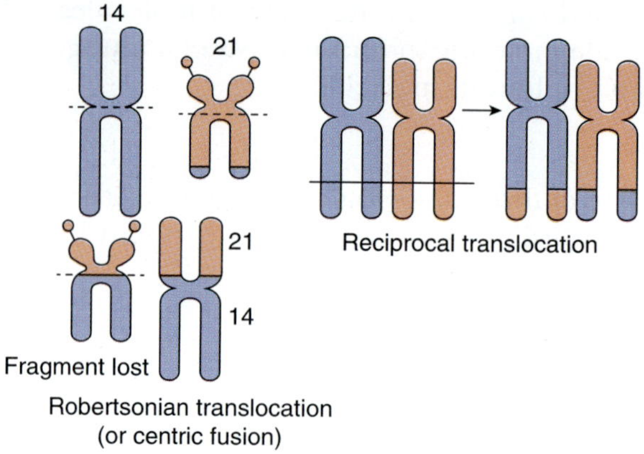

Figure 4.4 Types of translocations.

3. A condition in which the number of chromosome is not exact multiple of "*n*", i.e. 23, e.g. 45, X; 47, XX + 21; 47, XXY, etc.
4. Monosomy involving X chromosome is the only monosomy that is compatible with life.
5. There are two types of translocations. They are (*i*) Robertsonian translocation or centric fusion and (*ii*) reciprocal translocation (refer Fig. 4.4).
6. Also called "centric fusion"; it involves two acrocentric chromosomes, for example D/G translocation. The short arm of D group chromosome (13–15) and that of G group (21) fuse and fused fragment is lost. This is found in 4% of Down syndrome cases.
7. In this, there is an exchange of chromosome material distal to breaks and it involves non-homologous chromosomes.
8. This is because reciprocal translocation amounts to balanced translocation, there is no loss of chromosome material and hence these individuals have normal phenotype.
9. Individuals with reciprocal translocation, though have normal phenotype, produce abnormal gametes; eventually, this results in spontaneous abortion or baby with congenital malformations.
10. The formation of an isochromosome involves an abnormal split along the centromere leading to separation of arms instead of chromatids, for example, i (Xq), i.e. isochromosome X (refer Fig. 4.6).
11. c
12. b
13. b
14. 1–c; 2–d; 3–a; 4–b
15. c
16. Chimaera is an individual having two or more genetically different cell populations derived from more than one zygote, e.g. dispermic chimaera, blood group chimaera.

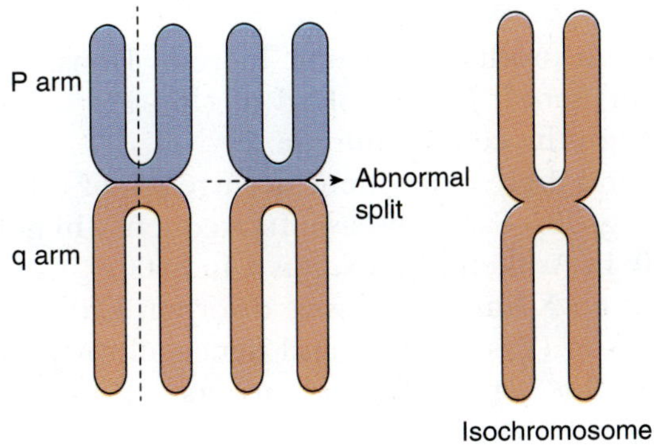

Figure 4.6 Formation of isochromosome.

Chapter 5: Developmental Genetics

1. They were first identified in Drosophila. The homeobox (*Hox*) genes determine segment identity in *Drosophila*. Four homeobox gene clusters have been identified in humans.
2. They are also called as paired box (*Pax*) genes. The paired box is a highly conserved DNA sequence that codes about 130 amino acids.

Chapter 6: Modes of Inheritance

1. Criteria for autosomal dominant trait are as follows:
   a. Affected person has an affected parent, exception being mutant gene.
   b. Affected person has normal and abnormal offsprings in equal proportion, i.e. there is 50% chance of dominant trait being transmitted to offsprings.
   c. Both males and females are equally affected.
   d. Trait appears in every generation without skipping.
2. In a dominant trait, usually one of the parents is affected; however, a normal parent can be expected in the following three situations:
   a. If the trait occurs because of a mutant gene.
   b. Gene, though present in the parent, has low expressivity.
   c. Extramarital paternity.
3. b
4. An autosomal recessive trait presents the following features:
   a. The trait appears in sibs and not in parents or offsprings.
   b. About 25% (1 in 4) of the sibs of the proband are affected.
   c. Both males and females have an equal chance of being affected.
   d. Parents of the proband may be consanguineous.
5. A heterozygote for an autosomal recessive trait is called carrier. Clinically the carrier appears to be normal. The carrier state is usually revealed by biochemical laboratory tests.
6. a

7. c
8. CFTR is a protein product of “Cystic Fibrosis Transmembrane Conductance Regulator gene”. It is involved in chloride transport and mucin secretion through the cell membrane.
9. A group in which there is frequent occurrence of otherwise rare recessive disorder, e.g. Tay–Sachs disease. Its frequency in general population is 1 in 360,000; in Ashkenazi Jews, it is 1 in 3600.
10. Haemophilia is an X-linked recessive disorder. Father passes on Y chromosome to his son (Y is normal) and hence all sons will be normal and all daughters of this person shall be carriers.
11. c
12. Located at X p 21, according to linkage analysis it is the largest gene identified so far, with 2300 kilo bases of genomic DNA.
13. They are deletion sites in DMD gene detected after Southern blot analysis with cDNA probes. The first is located in the first 20 exons of the gene while the other is in the centre of the gene between 45 and 53 exons.
14. It is unique in the sense that it is caused by a combination of a mutant gene with an associated cytogenetic abnormality.
15. When a single gene or gene pair produces multiple phenotypic effects, it is called pleiotropy, e.g. phenylketonuria. In this, an enzyme phenylalanine hydroxylase is deficient. It leads to multiple secondary effects such as mental retardation, phenylketonuria, hypopigmentation, etc.
16. Trait determined by autosomal transmission but expressed only in one sex is called sex-limited trait, e.g. precocious puberty affects males.
17. The trait is said to be sex-influenced, when it expresses in both sexes but with different frequencies, e.g. baldness, an autosomal dominant trait affecting males more frequently.
18. They are:
    a. Height/stature
    b. Intelligence quotient (IQ)
    c. Total ridge count (TRC)
    d. Cleft lip, cleft palate
19. It is a shorthand method of giving relevant information and also the mode of transmission of the disorder in the family.
20. 1–b; 2–c; 3–a; 4–d

Chapter 7: Biochemical Genetics

1. Inborn error of metabolism means a genetically determined enzyme defect leading to biochemical disorders, e.g. alkaptonuria, acatalasia, Hurler syndrome, etc.
2. PKU stands for phenylketonuria. Enzyme phenylalanine hydroxylase is deficient. It presents with blond hair, blue eyes, lack of pigmentation in brain, e.g. in substantia nigra. There is mental retardation.
3. Haemoglobinopathies are a group of disorders of haemoglobins, e.g. sickle cell disease, thalassaemias.

4. A branch of genetics that deals with genetically determined variations that become evident with an altered response to drugs.
5. It explains:
   a. Sensitivity of some individuals to a particular drug
   b. Defining potency of a drug
   c. Antibiotic resistance in some strains of bacteria
   d. Resistance of organophosphorus compounds in some insects
6. Vogel used the term pharmacogenetics for the first time about four decades ago.
7. Majority of drugs are metabolised through the following routes:
   a. Conjugation in liver, e.g. morphine, codeine are processed this way.
   b. Acetylation in liver, e.g. isoniazid, sulphonamides follow this route.
8. Detected in 1946 by Japanese clinician, it is deficiency of an enzyme catalase in red blood cells. Blood turns brownish black on exposure because haemoglobin is oxidised to methaemoglobin.
9. G6PD deficiency (glucose-6-phosphate-dehydrogenase) is inherited as X-linked recessive disorder.

Chapter 8: Genetics of Blood Groups

1. The ABO blood group system was discovered by Landsteiner, in 1900.
2. Red cell antigens are used as genetic markers in population studies because they satisfy following criteria:
   a. They form different phenotypes.
   b. They follow a simple pattern of inheritance.
   c. Their frequency is different in different populations.
   d. They are not influenced by environmental factors or age.
3. The presence or absence of A and B antigens gives rise to following four phenotypes: A, B, AB, and O.
4. The ABO locus is situated on the long arm of chromosome 9.
5. Gene "O" is amorph and has no effect. It is recessive to gene A and B, while both A and B genes are codominant.
6. Theoretically blood group "O" is universal donor, and group AB is assumed to be universal recipient.
7. Bombay phenotype or O_h phenotype was identified in Bombay in 1952. In this, red blood cells as well as secretions (e.g. saliva) lack antigens A, B, and H. The individuals with O_h phenotype are homozygous for gene "h", i.e. they have "hh", genotype. They do not form H substance, therefore no substrate is available for making antigen A and B, although A and B genes are present.
8. It is located on chromosome number 1, exhibiting eight alleles for five Rh antigens D, C, E, c and e.
9. Rh-null blood group persons have no Rh antigen on their red blood cells. The Rh antigen forms important and integral part of the red cell membrane, lack of this antigen causes haemolysis and eventually haemolytic anaemia.

10. In this, foetal RBCs die due to the action of antibodies formed by the mother (Rh-negative) against foetal Rh antigen. It begins in utero but continues after birth for about 3 months, the time taken by the maternal antibodies to get cleared from newborn circulation.

Chapter 9: Immunogenetics

1. It deals with the genetic basis of the immunological phenomenon in an organism/individual.
2. It refers to immunity conferred by the formation of antibodies produced by B cells (Bursa dependent cells).
3. It is immune response of an individual with the first exposure to an antigen. It is usually slow and takes relatively longer period.
4. These are serum proteins produced by plasma cells (activated B cell) in response to antigen. They are also called as antibodies.
5. An immunoglobulin molecule consists of four polypeptide chains. Two identical light (L) chains and two identical heavy (H) chains. These are held together by disulphide bonds.
6. This is because V (variable) and C (constant) region of each chain are coded by different genes.
7. On exposure to an antigen, B cell produces IgM antibody initially. However, on subsequent exposure to the same antigen it produces IgA or IgG antibody, still retaining specificity to the same antigen. This is labelled as class-switching.
8. They are as under:
   a. Autograft: Graft of the host's own tissue
   b. Isograft: Graft from genetically identical person, e.g. monozygotic twin
   c. Allograft: Graft from genetically non-identical person
   d. Xenograft: Graft between different species
9. An autograft and isograft have better acceptability because of the histocompatibility.
10. The histocompatibility means antigenic similarity between the donor and the recipient.
11. They are (*i*) mixed lymphocyte culture test and (*ii*) lymphocyte cytotoxicity test.
12. Also called human leucocyte antigen (HLA) system, it consists of four closely linked loci associated with the short arm of the chromosome 6. These are D (and D-related/DR), B, C and A.
13. An H-Y antigen is located on chromosome 6. It is governed by regulatory gene on Y chromosome. The H-Y antigen gene is thus a Y-linked histocompatibility gene.
14. It refers to a group of serum proteins required to inactivate foreign material after formation of antigen–antibody complex.
15. There are two types of immune responses:
    a. Cellular immunity—conferred by T-cell, these are thymus dependent cells.

b. Humoral immunity—conferred by formation of antibodies produced by B-cells, these are bursa-dependent cells. Bursa of Fabricius is an organ in birds responsible for formation of these cells.

16. It is the substance that evokes immune response.
17. After the first exposure, the subsequent exposure to the same antigen evokes a rapid and a more pronounced, so-called secondary response. This is possible through "memory cells".
18. They are lymphocytes that are primed to act vigorously with re-exposure to the same antigen.
19. The prime factor responsible for diversity in immunoglobulins (antibodies) appears to be multiple combinations of heavy and light chains.
20. 1–b; 2–c; 3–e; 4–a; 5–d.
21. They are kappa (κ) light chain and lambda (λ) chains.
22. Genes coding for polypeptide chains of immunoglobulins are associated with the following autosomes.
    a. H: Heavy chain on chromosome 14
    b. κ: Kappa light chain on chromosome 2
    c. λ: Lambda light chain on chromosome 22
23. Hybridoma can be obtained by fusing two cell components:
    a. The mutant mouse myeloma cells (not capable to produce antibody).
    b. Any cells capable of producing antibody, e.g. normal spleen cells from mice immunised with specific antigen or human lymphoblastoid cells. The hybrid cells thus produced can synthesise a single type of Ig/immunoglobulin, i.e. monoclonal antibody.
24. SCID stands for severe combined immunodeficiency disease. It is autosomal recessive trait owing to the lack of ADA (adenosine-deaminase).

Chapter 10: Cancer Genetics

1. Normal mammalian cells contain DNA sequences homologues to viral oncogenes; these are called proto-oncogenes or cellular oncogenes.
2. The retroviral oncogenes are formed because of errors in the replication of retroviral genome, following their integration at random sites in the host DNA. The resultant being viral oncogene (*V-onc*).
3. The anti-oncogenes are often called "tumour suppressor genes". These are the genes that suppress tumorigenic activity, e.g. Rb gene (retinoblastoma gene) is recessive tumour suppressor gene. Absence of the gene product in homozygous state leads to development of retinoblastoma.
4. PDGF is platelet derived growth factor. It stimulates proliferation of the connective tissue cells. EGF stands for epidermal growth factor; it promotes proliferation of epidermal as well as numerous other cells.
5. In cancer, the following chromosomal aberrations are found:
    a. Translocations, e.g. CML—Chronic myeloid leukaemia: Philadelphia chromosome [22q-(ph^1)]
    b. Deletions, e.g. Retinoblastoma (13q-), Aniridia-Wilms tumour (11p-)
    c. Chromosomal breakage syndromes, e.g. xeroderma pigmentosum

Chapter 11: Genetic Component in Common Diseases

1. Some individuals are more prone to develop common diseases like hypertension, diabetes, stroke, etc. This is because they have inherited predisposition or genetic susceptibility for that particular disease.
2. It may be caused by single gene defect leading to abnormal gene product that in turn alters metabolism.
3. The approaches available for determining genetic susceptibility are: (*i*) family study, (*ii*) twin study, (*iii*) population study, (*iv*) immigration study, (*v*) animal models, etc.
4. LDL receptors are glycoproteins that help the passage of low-density-lipoprotein (LDL) particle into the cell.
5. LDL-receptor gene is located on chromosome 19.
6. IDDM stands for insulin-dependent diabetes mellitus.
7. Body mass index is a parameter to gauge obesity. BMI $= W/H^2$; where W is weight in kilograms, and H is height in metres. A person with BMI more than 30 is said to obese. BMI between 25 and 30 denotes overweight.

Chapter 12: Population Genetics

1. It deals with study of genes in population.
2. The law states that "gene frequencies in a population remain constant from generation to generation, if no evolutionary factors such as migration, mutation, selection and drift are operating.
3. Eugenics refers to establishing the best characters in population by selective breeding.
4. It means writing on the skin. It includes study of ridge patterns on skin of palms, digits and soles.
5. It is a single transverse crease on palm, found in nearly 50% of Down syndrome patients.

Chapter 13: Prenatal Diagnosis

1. In the following cases, prenatal diagnosis is must:
   a. Maternal age above 35/40 years.
   b. If one of the parent is a balanced translocation carrier.
   c. If the couple already has a baby with neural tube defect or chromosomal abnormality.
   d. In a pregnancy at high risk for an autosomal or X-linked recessive disorder, which is severe.
2. It is a prenatal diagnostic procedure in which amniotic fluid is tapped and studied.
3. Amniocentesis is performed between 14 and 16 weeks because around that time sufficient amount of amniotic fluid is available for tapping, without harming the conceptus.
4. The procedure carries 1%–1.5% risk of abortion; however, a repeat amniocentesis increases the risk to about 9%.

5. They are:
   a. Increase in neonatal respiratory distress.
   b. Increase in postural orthopaedic problems such as talipes equinovarus, congenital dislocation of hip.
   c. Increased risk of rhesus sensitisation and perinatal mortality.
6. a. The chorion villous sampling (CVS) can be done at much earlier stage (around 8–10 weeks).
   b. Faster results because direct chromosome analysis can be done without culture.
   c. If the results indicate abnormality, termination of pregnancy can be safer and simpler in the first trimester.
7. There are (*i*) maternal cell contamination, and (*ii*) pseudomosaicism. It may be due to:
   a. An abnormal cell line appearing as an artefact during culture.
   b. Maternal cell contamination.
   c. Cell line from extra embryonic tissue.

Chapter 14: Genetic Counselling

1. It is defined as a process in which patient/s or their relatives at risk of a genetic disorder are made aware of the consequences of the disorder, its transmission and the ways by which this can be prevented or mitigated.
2. The steps are as under:
   a. An accurate diagnosis of the disorder.
   b. Advising patient and relatives after screening family members/relative with similar complaints for disease confirmation or otherwise normal relatives for carrier detection.
   c. Management of the disease curative or supportive.

Chapter 15: Gene Therapy

1. They are (*i*) liposome-mediated DNA transfer and (*ii*) receptor-mediated endocytosis.
2. The first human gene therapy was instituted on 14 September 1990 to a 4-year-old girl suffering from adenosine-deaminase (ADA) deficiency leading to severe combined immunodeficiency (SCID).

Chapter 16: Stem Cell Therapy

1. They are totipotent/pluripotent cells having endless self-renewal potential.
2. The sources are (*i*) embryonic stem cells (ESCs)—blastocyst; (*ii*) adult stem cells—PBSCs, BMSCs.
3. The stem cell therapy can be useful in stroke, brain injury, Alzheimer disease, myocardial infarction, diabetes, muscular dystrophy.
4. The research on embryonic stem cells (ESCs) raises the following issues:
   a. When does human life begin?
   b. Does the use of zygote for research account for killing?

Chapter 17: Epigenetics

1. Epigenetics is defined as heritable phenotype resulting from changes in chromosome without alterations in the DNA sequence.
2. The two basic epigenetic mechanisms involving DNA are:
   a. DNA methylation
   b. Histone modifications
3. At the 5′ end of the first exon, there is high concentration of CG sequences. Cytosine in these gets methylated forming 5′ methylcytosine. Majority of them are found in CpG islands (cytosine–phosphate–guanine). Methylation of these CpG islands is associated with *silencing* of genes (imprinting).
4. Genomic imprinting can be of paternal or maternal alleles, i.e. those alleles/genes are silenced, e.g.
   - Maternal imprinting—Prader–Willi syndrome
   - Paternal imprinting—Angelman syndrome

Bibliography

1. Anderson WF. Prospects for human gene therapy. *Science* 1984; 226: 401–9.
2. Anderson WF. Human gene therapy. *Science* 1992; 256:808–13.
3. Annual Review of Genetics. California, USA: Annual Review Inc., 1985.
4. Benson PF, Fenson AH. Genetic biochemical disorders. Oxford: Oxford University Press, 1985.
5. Borgaonkar DS. *Chromosomal Variation in Man: A Catalogue of Chromosomal Variants and Anomalies.* 4th ed. New York: AR Liss, 1984.
6. Brock DJH, Mayo O. The Biochemical Genetics of Man. 2nd ed. London: Academic Press, 1979.
7. Brock DJH, Rodeck CH, Ferguson Smith MA. *Prenatal Diagnosis and Screening.* Edinburgh: Churchill Livingstone, 1992.
8. Bundey S, Brett EM, Emery AEH. *Genetics and Neurology.* 2nd ed. Churchill Livingstone, 1992.
9. Cappecchi MR. Targeted gene replacement. *Scientific American* 1994; 270:34–41.
10. Ciba Foundation Symposium. Human genetics: Possibilities and realities. Amsterdam: Excerpta Medica, 1979.
11. Conner JM, Ferguson S. *Essential Medical Genetics.* 2nd ed. Oxford: Blackwell Scientific, 1987.
12. Darnell J, Lodish H, Baltimore D. *Molecular Cellular Biology.* New York: Freeman, 1990.
13. Emery AEH. Portraits in medical genetics. *J Med Genet* 1989; 26:116–8.
14. Emery AEH. Duchenne Muscular Dystrophy. 2nd ed. Oxford: Oxford University Press, 1993.
15. Emery AEH, Rimoin DL. *Principles and Practices of Medical Genetics,* Volumes 1 and 2. Edinburgh: Churchill Livingstone, 1983.
16. Fincham JRS. Genetics. John Wright, 1983.
17. Fraser G, Mayo O. *Textbook of Human Genetics.* Oxford: Blackwell Scientific, 1975.
18. Fraser RJ, Pembrey ME. *Introduction to Medical Genetics.* 8th ed. Oxford: Oxford University Press, 1985.
19. Goodenough U. *Genetics.* 3rd ed. Japan: Saunders, 1984.
20. Harper P. *Practical Genetic Counselling.* Oxford: Butterworth Heinemann, 1993.
21. Stone R. *Harrison's Principles of Internal Medicine.* McGraw Hill, 1987.
22. Hodgson SV, Maher ER. *A Practical Guide to Human Cancer Genetics.* Cambridge University Press, 1993.
23. Jorde Lynn B. *Medical Genetics.* Mosby, 2005.
24. King RC, Stansfield WD. *Dictionary of Genetics.* 3rd ed. Oxford: Oxford University Press, 1985.

25. Lachmann PJ, Peter DK. *Clinical Aspects of Immunology*. 4th ed. Oxford: Blackwell Scientific, 1992.
26. Lewin B. *Genes V*. Oxford University, 1994.
27. Lilford RJ. *Prenatal Diagnosis and Prognosis*. London: Butterworth, 1990.
28. Mckusick VA. *Mendelian Inheritance in Man*. 11th ed. Baltimore. John Hopkins University Press, 1994.
29. Pritchard DJ. *Foundations of Developmental Genetics*. London: Taylor and Francis, 1986.
30. Shubha R Phadke. *Genetics for Clinicians*. Prism Books Pvt. Ltd., 2007.
31. Swanson CP, Merz T, Young WJ. *Cytogenetics*. 2nd ed. Englewood Cliffs: Prentice-Hall, 1981.
32. Thompson JS, Thompson MW. *Genetics in Medicine*. 7th ed. Philadelphia: WB Saunders, 2006.
33. Tumpenny P, Ellard S. *Emery's Elements of Medical Genetics*. 12th ed. Churchill Livingstone, 2004.
34. Vogel F, Motulsky AG. *Human Genetics: Problems and Approaches*. 2nd ed. Berlin: Springer–Verlag, 1986.
35. Weatherall DJ. *The New Genetics and Clinical Practice*. 3rd ed. Oxford: Oxford Medical Publications, 1991.

Index